EDITOR: MARTIN WIND

OSPREY
MILITARY

ELITE SERIES

6

FRENCH FOREIGN LEGION PARATROOPS

Text by
MARTIN WINDROW
and WAYNE BRABY
Colour plates by
KEVIN LYLES

Published in 1985 by
Osprey Publishing Ltd
59 Grosvenor Street, London W1X 9DA
© Copyright 1985 Osprey Publishing Ltd
Reprinted 1986 (twice), 1987, 1988, 1991

Author's note on 1986 reprint edition:
Where this text differs in matters of uniform detail
from data in my book *Uniforms of the French Foreign
Legion 1831–1981* (Blandford Press), this more recent
research should be regarded as more correct.

A number of regrettable errors in the original 1985
edition of this text, mainly involving the designations
of different early models of French parachute
uniforms, are corrected in this reprint.

It is hoped to follow this title with a future book in
the Elite series which will cover Foreign Legion
ground troops—infantry, cavalry, Saharan and
engineer units—over the same period.

MCW

British Library Cataloguing in Publication Data

Windrow, Martin and Braby, Wayne
  French Foreign Legion Paratroops.—(Elite; v.6)
  1. France. *Armée. Légion étrangère*—History
  2. France. *Armée*—Parachute troops—History
  I. Title   II. Series
  356′.166′0944   UA703.L5

  ISBN 0-85045-629-0

Filmset in Great Britain
Printed in Hong Kong

Acknowledgements
The authors and illustrator wish to record their
thanks to Simon Dunstan, Rob Flynn, William
Fowler, Sam Katz, Mike Miles and Lee Russell for
their help with the preparation of illustrations; and to
Christa Hook for the line and tone drawings. We also
gladly acknowledge our debt to the published
researches of M. Denis Lassus.

French Foreign Legion Paratroops

The Indochina War 1946–54

The garrisons of French Indochina (today Vietnam, Laos and Cambodia) spent much of the Second World War in enforced idleness, supervised by Japanese forces whose presence had been sanctioned by the Vichy government in October 1940. In March 1945 the Japanese turned without warning on the scattered and ill-equipped French garrisons: those who resisted were massacred with characteristic savagery. A heroic column, whose major element was drawn from the Foreign Legion's 5th Infantry Regiment (5ᵉREI) fought its way north into Nationalist China, where it was interned. It was men of the 5ᵉREI who became the first légionnaire-paras of all when, at Kun Ming USAAF base, Lt. Chenel's platoon began jump training shortly before Japan's surrender.

The power vacuum in Indochina was skilfully exploited by a Vietnamese nationalist organisation dominated by the Communist leadership of Ho Chi Minh and his military subordinate Vo Nguyen Giap. It was February 1946 before war-torn France was able to send troops to Hanoi. By that time Ho had used the respite offered by a sketchy inter-Allied occupation to turn his 'Viet Minh' into a well-armed nationwide guerrilla organisation, and to declare an independent republic. The talks which followed were doomed from the start. The Viet Minh had no intention of accepting the re-imposition of French colonial rule; and the French, still smarting from the traumas of the Second World War, were in no mood for compromise. Negotiations broke down, accompanied by violent episodes which left each side with a legacy of hatred.

In December 1946 the Viet Minh took to the hills; and early French attempts to destroy their refuges among the wooded cliffs of Tonkin, north of the French centres of Hanoi and Haiphong in the Red River Delta, met with failure. Ho and Giap

Patrice de Carfort, *médecin-lieutenant* of the 1ᵉʳBEP during 1953. He wears a British 1942 windproof smock and trousers, the former with a front zip fastener added.

activated their organisation throughout the peninsula; and over the next eight years they tied down, and eventually defeated, the French Union forces in the world's first text-book demonstration of Maoist revolutionary warfare.

The war was fought, on the one side, by a Viet Minh army which by early 1951 could field 100,000 to 120,000 regular soldiers grouped in one artillery and five infantry divisions of conventional size and organisation, and in a number of independent regimental groups. This regular *Chuc Luc* was backed by a strong regional militia of perhaps twice that strength; and enjoyed the active or passive

The gates of Bas-Fort St Nicolas in Marseilles—the first step towards the hills and paddy fields of Indochina for tens of thousands of 'men with nothing to lose' in the years 1945–54.

support of a significant part of the population. While France's fighting men had the healthiest respect for the 'Viets', her high command persisted for some years in regarding them as mere colonial bandits. Even though the revelation of Giap's conventional strength in autumn 1950 opened French eyes, a fatal degree of underestimation of the enemy's abilities was still to have dire effects as late as 1954.

On the other side was an Expeditionary Corps whose strength varied between about 115,000 and 185,000 men. Of this total, only some 50,000 were French. Legally, French conscripts could not be sent to Indochina; thousands of Frenchmen served as cadres with African and Asian units, but the only formed, all-French units available were volunteer Colonial regiments, and specialist units of regular career soldiers such as airborne and armoured regiments. North and West African colonies provided some 25,000 men, mainly infantry. At various times between 50,000 and 90,000 locally recruited Vietnamese and *montagnards* provided the largest single element. They were generally lightly

equipped, and most of them were tied down in static defensive posts along the highways and waterways and in the wild interior. From 1950 onwards an intense effort was made to form a serious Vietnamese National Army, with variable results. Finally, some 20,000 men at any one time were provided by the French Foreign Legion. In the chaotic aftermath of world war, it was not surprising that the Legion was swollen to an unprecedented strength of around 30,000 by the lost ex-soldiers and refugees of 50 nations. (The German element was strong, but has been exaggerated by some journalists.)

The general nature of the war should be sketched in, however briefly. The Viet Minh enjoyed priceless advantages from the start; and the French suffered under severe handicaps.

The key element, as in all revolutionary wars, was the support enjoyed by the VM among the population. Giap received a constant flow of information about French movements, and could orchestrate the activity of his regular and militia units to take advantage of them. The use of inappropriate, conventional-warfare methods against an elusive enemy hardened this popular hostility towards the French: there is no quicker way of turning a neutral bystander into an active opponent than by bombing his village or destroying his crops.

The terrain was perfect for all kinds of irregular operation. A virtually unexplored hinterland of thickly jungled mountains was surrounded by swampy plains, and traversed by only a few narrow and winding roads. The Viet Minh nearly always had the initiative, and the 'interior lines'. They could achieve massive local superiority for widely separated attacks on lonely posts; and could ambush the French reaction forces—tied as they were to a few, predictable roads—before fading back into the wilderness.

During 1946–49 the VM fought a limited war of guerrilla pinpricks, inflicting a steady drain on French manpower while building up their own conventional forces. From the outset, the French truly controlled only the immediate vicinity of their main towns and routes; and after dark there was no such thing as a 'safe road' in the whole country.

Giap unveiled the potential of his regular divisions in the autumn of 1950, when he captured a large area of north-eastern Tonkin along the Chinese border in the campaign generally known as 'the battle of RC 4'. Thereafter he enjoyed the major advantage of 'active sanctuary' inside China, which had fallen to his Communist sponsors in 1949. With training camps, supply dumps and rest areas safe on the other side of a mountainous jungle frontier, and with an increasing flow of the most modern materiel, he was able to pursue ever more ambitious campaigns of movement. Once, the French had prayed that the Viet Minh would come down from the hills so that they could be destroyed by artillery, tanks and aircraft. But in the event, France's always-inadequate tactical airpower did not prove decisive: the VM were masters of camouflage and of concealed movement—even by major formations, huge columns of supply coolies,

Légionnaires of the 2ᵉBEP in Indochina, August 1953. The sergent-chef at left, wearing the fabric bush hat, displays (unusually) his three gold rank chevrons and Legion écusson on the sleeve of his US camouflage jacket. The biblically-bearded stalwart on the right—a pioneer?—wears the locally modified US M1 helmet, with fabric chin strapping. Note also his French webbing suspenders, and belt with British '37-style buckle.

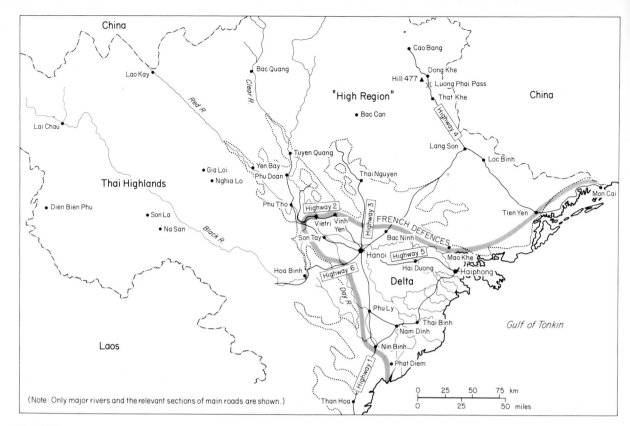

The map shows locations including: China, Lao Kay, Bac Quang, Cao Bang, Dong Khe, Hill 477, Luong Phai Pass, That Khe, "High Region", Bac Can, Lang Son, Loc Binh, Lai Chau, Red R, Clear R, Tuyen Quang, Gia Loi, Yen Bay, Phu Doan, Nghia Lo, Thai Nguyen, Thai Highlands, Phu Tho, Vietri, Vinh Yen, Highway 2, Highway 3, FRENCH DEFENCES, Tien Yen, Mon Cai, Dien Bien Phu, Son La, Na San, Black R, Son Tay, Bac Ninh, Hanoi, Highway 5, Mao Khe, Hai Duong, Haiphong, Hoa Binh, Highway 6, Delta, Day R, Phu Ly, Gulf of Tonkin, Thai Binh, Nam Dinh, Nin Binh, Phat Diem, Laos, Highway 1, Than Hoa

(Note: Only major rivers and the relevant sections of main roads are shown.)

0   25   50   75 km
0   25   50 miles

North Vietnam was the cockpit for the most serious fighting of 1950–54: the BEPs' major battlefields are marked on this simplified map. The major Viet Minh strongholds lay in the 'Viet Bac', around, east and south of Bac Can.

and strong artillery assets. French armour and artillery were limited to the roads and the few areas of 'good going'; and while they did achieve some notable victories, the French high command never had sufficient mobile reserves to exploit victory decisively. Airborne units were employed effectively as 'quick reaction' forces; but as they were often sent into action to retrieve losing battles, they inevitably suffered very high casualties. Inadequate airlift capacity was a constant problem: at the beginning of 1949 there were only 45 old Junkers Ju52 trimotor transports available, and even at the end of the war in 1954 France had only some 80 C-47 Dakotas in-country.

A feature of the war was the Viet Minh's exploitation of the difference between European and Asian attitudes. When they attacked a post or ambushed a convoy, they knew that motorised or airborne reserves would always try to fight their way to the rescue. Their tactical plans always took account of this, and the luring of such rescue parties

into massive ambushes became a commonplace of the war. By contrast, the VM never reinforced defeat: units which got into difficulties were ruthlessly abandoned. The same sacrificial attitude was observed during their major attacks on French positions. Intelligence-gathering, planning, and painstaking rehearsal to the highest standards of conventional armies were followed by suicidal 'human wave' assaults: a combination of professional excellence and inhuman motivation which the French found unnerving.

An equally impressive aspect of the war at unit level was the self-sacrificing courage and endurance of many of the French Union troops. High in this company stand the battalions of the Foreign Legion, who left more than 10,000 dead *képis blancs* in Indochina; and few stand higher than the 1st and 2nd Foreign Parachute Battalions.

Légionnaires with wings
The decision, in spring 1948, to form two battalions of Foreign Legion paratroopers was prompted by the requirement for enlarged airborne forces in Indochina, and the healthy recruitment then enjoyed by the Legion. There were some initial

doubts. The Legion were known to be magnificent heavy infantry, 'genre rouleau compresseur' ('steam roller types'); but were felt by some to lack the flexibility and agility demanded by independent airborne operations. In the Legion itself there were some misgivings over the possible clash between the self-consciously exclusive 'para mentality', and the Legion's own very marked ésprit de corps. All these doubts evaporated with experience.

The first operational Legion para unit was formed at Hanoi on 1 April 1948 from platoons of the 2ᵉ and 3ᵉREI and the 13ᵉ Demi-Brigade—Legion infantry regiments already operational in-country. This 'Compagnie Parachutiste du 3ᵉREI' was commanded by Lt. Morin of II/3ᵉREI (2nd Bn., 3rd Foreign Infantry); like a number of other prominent officers of Legion paras, Morin was a survivor of the Nazi concentration camps. The CP 3ᵉREI was attached for administrative and operational purposes to III/1ᵉʳRCP (3rd Bn., 1st Regt. of Parachute Chasseurs—a French Metropolitan unit). Its first operation, at Van Xa, took place on 26 April; its first firefight, on 5 May at Aymo Tong. For the next year the company operated all over northern Tonkin, fighting on RC 3 and 4 ('Route Coloniale'—'Colonial Highway') and around Cao Bang, Lang Son, and other important French outposts.

Meanwhile, in French North Africa, formation of two complete battalions proceeded. On 13 May 1948 a Groupement d'Instruction Parachutiste was installed at Khamisis near Sidi bel Abbès, the Legion's Algerian depot town. On 1 July 1948 the 1ᵉʳ Bataillon Étranger de Parachutistes was officially formed under the command of Capt. Pierre Segrétain; and between July and October jump training was carried out at Philippeville. The 2ᵉBEP, largely formed from personnel of Legion units in Morocco, was installed at Sétif, Algeria, from 1 October 1948. (The first man officially posted to the 2ᵉBEP was Lt. Caillaud, who would command the 2ᵉREP in 1963–65 during its transition into a para-commando regiment.) By December 1948 the 2ᵉBEP was completing its jump training. Few of the first intake of officers of either battalion were légionnaires; the majority came from the Metropolitan and Colonial para units of the 25th Airborne Division, as did a cadre of NCOs. The men were volunteers from Legion infantry and

French paras emplane for a dawn jump from C-47s, early 1950s. Most combat jumps involved insertion on to extremely marginal DZs deep in 'Indian country' and far from any hope of support, against an enemy in unknown but often overwhelming strength. The right hand man seems to carry an FM 24/29 in a quick-release valise. (ECPA)

cavalry units; many had already completed a two-year combat tour in Indochina, and some had parachute experience in other armies.

On 20 October 1948 the 1ᵉʳBEP sailed for Indochina, landing at Haiphong, North Vietnam on 12 November. The 2ᵉBEP sailed on 13 January 1949, landing at Saigon on 8 February. For reasons of space, the combat history of these units follows in note form; and only major, or particularly characteristic operations are listed. The reader should bear in mind that gaps in the chronology do not indicate periods of idleness: French airborne units were heavily committed to security operations at those times when their intervention as reserve elements was not required.

On 16 November 1949 the 3ᵉBEP was formed from personnel of the 1ᵉʳREI, and based at Sétif. This was a training and transit unit, providing replacements for the two battalions in Indochina.

Operations, 1ᵉʳBEP, Nov. 1948–Nov. 1953:

15 Nov. 1948 1st Co. posted Haiduong; remainder, still largely unarmed, by train to Hanoi—suffered first casualties to mine explosion during journey.

24 Nov. 2nd Co. detached Lang Son, important 'anchor' of posts along RC 4, highway following Chinese frontier in NE Tonkin. Already frequent heavy ambushes of once- or twice-weekly convoys up RC 4 to Cao Bang, most advanced post in High Region; worst stretch c. 70 km That Khe-Cao Bang. During following months, 2nd Co. rotated platoons

Often captioned as Dien Bien Phu, this shot in fact shows a 1ᵉʳBEP counter-attack getting under way during the defence of the entrenched camp at Na San in November 1952. (ECPA)

as convoy escorts. March 1949, co. moved That Khe, attached II/3ᵉREI during period almost daily combats. 25 April 1949, heavy fighting during ambush Loung Phai pass. Meanwhile:

8 Dec. 1948 Remainder bn. fought first combat Xuan Mai, between Hanoi and Hoa Binh: VM co. wiped out for light losses.

18 Mar. 1949 Operation 'Bayard'—first operational jump: successful op. NE of Haiphong.

8 May 2nd Co. rejoined bn.

12 May Op. 'Pomone'—bn. re-took Tuyen Quang, scene of famous Legion battle 1885, but soon withdrawn.

31 May CP 3ᵉREI disbanded, personnel to 1ᵉʳBEP.

1 July First op. by complete bn.

Mid-Oct. Op. 'Thérèse'—bn. jumped near Loung Phai pass, RC 4 after ambush Spahi unit 13 Oct. and radio silence Legion garrison Loung Phai post. After 'blind' jump over appalling DZ among wooded cliffs, found burnt-out trucks and dead of ambush; but Loung Phai garrison found still holding out.

27 Nov. Report 5ᵉREI platoon lost near Cho Bo post on Black River. HQ Co., 2nd Co. jumped Hoa Binh, marched 25 km to Cho Bo—still holding out, but nearby Suyut post found massacred, and 5ᵉREI platoon confirmed wiped out.

22 Dec. 1949–15 Jan. 1950 Op. 'Diabolo'—security ops., Red River Delta.

8 Feb. 1950 Op. 'Tonneau'—successful forced march to re-take Thai Binh.

Disaster on RC 4:

Sept. 1950 Warned major VM build-up for offensive N. Tonkin at end of rains, French high command decided to abandon indefensible line of posts along RC 4; but caught out by Giap's opening moves mid-Sept., not anticipated for another month.

16–17 Sept. Vital post Dong Khe, held by two cos. II/3ᵉREI, wiped out in 48 hrs close combat by five bns. of VM Regts. 165, 174 with artillery support.

17–18 Sept. 1ᵉʳBEP jumped That Khe.

23 Sept. 1ᵉʳBEP probed NE towards Poma; narrowly fought way back That Khe, with intelligence suggesting (accurately) presence at least 15 VM bns. incl. two of artillery.

24 Sept. Col. Charton, OC Cao Bang, ordered prepare withdrawal 3 Oct.

1 Oct. 'Groupement Bayard' formed That Khe: 1ᵉʳBEP, three bns. Moroccan infantry. Left That Khe up RC 4 towards Dong Khe, with mission re-take Dong Khe. Further mission, to meet and escort Charton Column withdrawing from Cao Bang, not specified before departure. Heavy rain and mist prevented air support (from dozen Bell Kingcobras which represented entire available tac-air in whole of Tonkin).

2–4 Oct. 'Bayard' blocked, surrounded, split up, heavily attacked at Na Keo and Hill 615, S of Dong Khe, by major VM forces. Night 2/3 Oct., confirmation received Charton Column leaving Cao Bang. Night 3/4 Oct., signal received Charton also hopelessly blocked on RC 4, and taking to hill tracks to W, via Hill 477. Early hours 4 Oct., 'Bayard' began attempt disengage, move across jungle ridges to W, intending RV with Charton near Hill 477. (By now, Giap had up to 30 VM bns. from e.g. Regts. 36, 88, 99, 165, 174, 175, 209 and 246, plus artillery, assembling in area to exploit dislocation of French columns.)

4–7 Oct. Both columns fought way through hills over bad jungle tracks at agonisingly slow pace, frequently attacked. 'Bayard' seriously short of food, water, ammunition; support weapons virtually silenced; terrain forced back-carrying increasing numbers wounded. Ch. de Bn. Segrétain, 1ᵉʳBEP, seriously ill.

7 Oct. Very costly action to break encirclement in natural cliff amphitheatre near Coc Xa; 1ᵉʳBEP and Moroccan survivors eventually escaped down lianas hanging over face of sheer 75-ft cliff, many with wounded comrades lashed on backs. 1ᵉʳBEP down to c. 130 effectives. Many men separated, lost in jungle as 'Bayard' column broken up by firefights in close cover. Evening 7 Oct., 'Bayard' survivors RV'd with Charton near Hill 477—found Charton equally mauled. Constant enemy pressure. Commanders decided disperse force in platoon-size parties each led by officer or NCO with map, compass, and attempt infiltrate enemy-infested jungle towards RC 4, re-assemble That Khe. Wounded abandoned with volunteer medics. Segrétain, seriously wounded in ambush, ordered his men to leave him, and died shortly afterwards in captivity.

8 Oct. French high command attempted to reinforce crumbling RC 4 line by dropping 3ᵉBCCP (3rd Colonial Para-Commando Bn.), and company newly arrived 1ᵉʳBEP replacements led by Lt. Loth, at That Khe.

10 Oct. Capt. Jeanpierre, 2ic 1ᵉʳBEP, led 29 légionnaire-paras into That Khe; safely evacuated same day. These were only survivors, of 499 who left That Khe on 1 Oct., to reach French lines. Later on 10 Oct., That Khe garrison decided to abandon post (while many 'Bayard' and 'Charton' stragglers still trying desperately to reach it). Retreat down RC 4 began in some disorder. 3ᵉBCCP and Loth Co. destroyed in major ambushes; of latter, only 12 légionnaire-paras reached safety.

17–18 Oct. With major VM attacks throughout sector, and garrisons falling back in confusion, the vital stronghold of Lang Son was abandoned, in scenes compared by some officers to France in June 1940. Important centres far from direct threat—e.g. Lao Kay, Hoa Binh—were also evacuated as defeatism and panic spread.

Giap had captured everything north of the Red River Delta, securing his access to the Chinese frontier. He had inflicted between 6,000 and 7,000 casualties, including some 4,800 dead and missing; and had captured some 1,300 tons of supplies, with enough artillery, crew-served weapons and small arms to equip a complete division. The Expeditionary Corps, and France, were profoundly

Para of the 2ᵉBEP's 2ᵉCIPLE, wearing an M1 helmet locally modified to take extra chin strapping of stitched fabric, and the British 1942 windproof smock. The photo was taken during Operation 'Brochet', September 1953, by which time significant numbers of Vietnamese were serving in both BEPs.

TABLE 1

1ᵉʳ and 2ᵉ Bataillons Étrangers de Parachutistes
Indochina, 1948–55

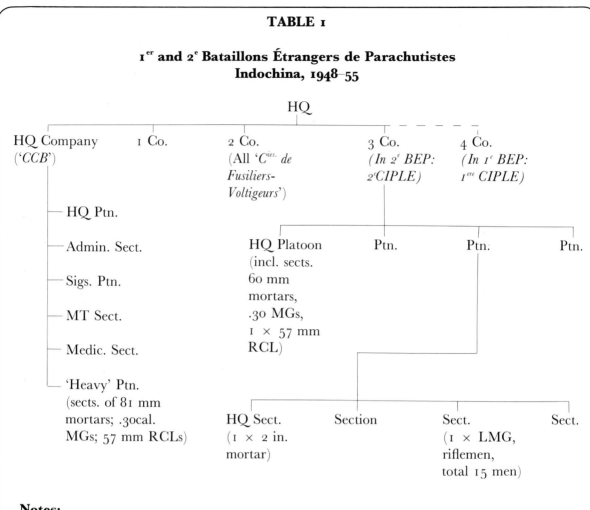

Notes:

Number, and numbering of companies varied; HQ plus three rifle cos. was normal. For a period the 2ᵉBEP numbered cos. '5, 6, 7, 8'. Numbering of CIPLEs also varied. Personnel strength fluctuated between c.400 and 1,000, incl. rear base staff.

shocked by this evidence of the enemy's strength and conventional military skills. The military and political authorities were now unified under command of Gen. Jean de Lattre de Tassigny, who arrived in Saigon on 17 December 1950 with full powers. He quickly improved morale; began construction of a vast chain of concrete forts around the Delta, to secure the French heartland; and pushed ahead with a programme of training local troops by integration with French units, looking forward to the day when a serious Vietnamese National Army could reinforce the Expeditionary Corps. In early 1951 Giap, made overconfident by his success, tried to push straight into the Delta with his regular divisions[1]. He was bloodily repulsed at Vinh Yen (January 1951), Mao Khe (March 1951) and on the Day River (May–June 1951). Meanwhile:

31 Dec. 1950 1ᵉʳBEP disbanded; survivors and rear base personnel formed 'Compagnie de Marche 1ᵉʳBEP' within 2ᵉBEP.

[1]In early 1951 Giap assembled his regiments into 10,000-man divisions. Each had three numbered regiments (e.g. 312 Div.—141, 165, 209 Regts.) plus an artillery battalion. Each regiment had a support element (typically, 4 × 75 mm guns, 4 × 120 mm mortars) plus four battalions. Each battalion had 20 MGs, 8 × 81 mm mortars, 3 × 75 mm RCLs, and several bazookas, in addition to small arms. Their ratio of SMGs to rifles would rise to one in three by 1954. American support weapons captured in Korea and passed on by China were more modern and plentiful than the Second World War vintage weapons of French battalions.

1 Mar. 1951 Bn. officially recreated; 3rd Co., 2ᵉBEP provided men for Cie. de M. 1ᵉʳBEP.

13 Mar. 400-plus replacements arrived Haiphong from North Africa, most from 3ᵉBEP; bn. organised with three companies.

18 Mar. 1ᵉʳBEP in action in northern Delta area.

Mar.–Nov. Continuous security ops. in Delta; typically, up to two months in field followed by two–three weeks Hanoi/Bach Mai rear base. May 1951, 1st Indochinese Para Co. Foreign Legion (1ᵉʳᵉCIPLE) formed, attached 1ᵉʳBEP, and Vietnamese ptns. added to other cos. in accordance De Lattre policy.

10 Nov. Op. 'Tulipe'—bn. jumped Cho Ben, preliminary objective Hoa Binh operation. (Last operation using Ju52 aircraft for combat drop.)

Hoa Binh, the Muong tribal capital 40 km south-west of the Delta defences, abandoned a year before, was now an important VM staging area, and De Lattre chose it for a major initiative. Taken without difficulty on 13 November by three para battalions, and garrisoned by five battalions, it turned into a costly liability. It was linked to the Delta by RC 6, more or less direct; and by a long northward loop of the Black River. Both routes were flanked by steep, wooded hills, perfect for ambushers. Giap quickly threw his 304, 308 and 312 Divs. into action, not against the entrenched camp (whose airstrip was constantly shelled, however) but against its lifelines. Heavy fighting along the Black River ended when the VM finally cut the river route on 12 January 1952; and massive attacks on posts and convoys finally cut RC 6 completely on 9 January. (By this time Gen. de Lattre, sick and disheartened, had returned to France, where he died on 11 January.) When the new C-in-C, Gen. Salan, decided to cut his losses and withdraw, it took a force of 12 battalions and three artillery groups with heavy air support a full 12 days to re-open RC 6 from the Delta end (18–29 January). The withdrawal of the Hoa Binh and intermediate garrisons, on 22–24 February, was one long running battle.

15 Nov.–mid Dec. 1951 1ᵉʳBEP fought repeated actions around Cho Ben, Ba Vi hills, Ap Da Chong and Notre Dame Rock during 'Black River' phase.

8–31 Jan. 1952 Bn. heavily engaged in 'RC 6' phase, particularly at Au Trach, Dong Ben.

Mar.–Oct. Bn. committed to successive security ops., southern Delta area.

In October 1952 Giap launched a major offensive south-westwards from northern Tonkin, passing west of the inland tip of the Delta defences, and deep into the Thai Highlands on the Laotian border in south-west Tonkin. This threatened pro-French Thai *montagnard* tribes, scattered French posts in the area, and Laos itself. It also drew overstretched French reserves into costly fighting in terrible terrain far from support; and in their absence, allowed heavy infiltration of the Delta itself. The important post of Nghia Lo fell, leaving Na San, south of the Black River, the only possible centre of resistance for retreating garrisons. A major airborne operation was now carried out, partly to buy time for Na San, partly to threaten the rear lines of Giap's forces:

9 Nov. Op. 'Marion', airborne phase Op. 'Lorraine'—30,000 French struck at VM rear depots around Phu Doan. 1ᵉʳ, 2ᵉBEP and 3ᵉBPC (3rd Colonial Para Bn.) jumped from 450 ft over rough DZ near Phu Doan; found and destroyed important dumps; RV'd with Groupes Mobiles 1 and 4, attacking up RC 2 from Vietri; withdrew safely down RC 2 on 16 Nov. (Road force, withdrawing 17 Nov., heavily ambushed.)

20 Nov. 1ᵉʳBEP flown into Na San; 20 km SE to Co Noi, where dug in as reception party for retreating garrisons—and pursuing VM 312 Div.

21/22 Nov. Withdrew to Na San perimeter in gruelling night march. 22 Nov., assigned eastern strongpoints ('PAs') 23, 23*bis*, 23*ter*.

23/24 Nov. Major enemy bombardment and infantry attacks. 2nd Co. and 1ᵉʳᵉCIPLE held PA 23*bis* with distinction.

24–30 Nov. Bn. assigned neighbouring PA 8, recaptured after loss previous night (1st and HQ Cos.), and PA 8*bis* 3rd Co.). Lull in infantry attacks; all PAs strengthened.

30 Nov./1 Dec. Heavy artillery and infantry attacks; PAs held.

1/2 Dec. Final 'human wave' attacks held with difficulty. VM 312 Div. left 600 dead on wire—but also evidence lavish Chinese supply of support and automatic weapons.

No further major infantry attacks; patrols probed area, found main VM force withdrawn. 1ᵉʳBEP sortied 9 and 15 Dec., some combat. 17 Jan., bn. returned Hanoi.

Late Jan.–Feb. 1953 Bn. ops. round Kontum, An

Khe, Ban Me Thuot in southern Annam.

1–4 Apr. Ops. Ha Dong, central Delta.

2–3 Aug. Bn. flown Tourane, central Annam.

15 Aug.–1 Sept. Ops. Plain of Jars, Laos.

1 Sept. 1st Foreign Para Heavy Mortar Co. (1^{ere}CEPML) raised from personnel both BEPs; attached 1^{er}BEP.

23 Sept.–11 Oct. Returned to Delta for Ops. 'Brochet I–IV', Le Khu area. Final result: 1^{er}BEP lost 96 casualties, VM lost 10.

21 Nov. 1953 Op. 'Castor'—bn. jumped over Dien Bien Phu.

Operations, 2^{e}BEP, Feb. 1949–March 1954:

Feb.–Nov. 1949 Bn. served as 'sector troops', Cambodia: dispersed cos. rotated for convoy escort, road opening, etc.

3 Nov. To Saigon/Tan Son Nhut as 'general reserve' unit, for emergency intervention Cochinchina, Annam, Cambodia, southern Laos. One co. on stand-by for immediate take-off; one on 12-hr alert; one on 24-hr alert; cos. rotated weekly.

26 Dec. First op. in this rôle. 1st Co. jumped over unprepared, unreconnoitred, partly flooded DZ at Tra Vinh, in high wind, at nightfall, into VM fire. Op. successful. (Six more combat jumps before Sept. 1950.)

2 Jan. 1950 Bn. moved Dong Hoi, central Annam.

6 Jan. First combat in bn. strength, Phuoc Long: enemy driven 10 km, 100 killed, for five dead, 25 wounded in BEP.

31 Mar. 130 enemy killed by 2nd Co. in hard combat, Ba Cun.

20 Sept. Moved Hanoi, 250 men of 1st, 2nd Cos. jumped Sin Ma Kay in hills N of Red River. Giap's RC 4 offensive (see under 1^{er}BEP) forced abandonment many posts W of Cao Bang; 2^{e}BEP dropped to help column Thai troops, families, Moroccan infantry falling back on Lao Kay. Muong guides led BEP to Column de Bazin, but two days lost gathering stragglers. Already out of food, force began march over appalling jungle-mountain tracks towards Lao Kay—30 km on map, much further on ground. Several firefights with pursuing VM; hot rearguard action while column rafted flooded Song Chay River. Reached Lao Kay after three-day march—2^{e}BEP marched into post singing, in column-of-threes. Flown Lao Kay–Hanoi,

where news of loss 1^{er}BEP arriving.

8 Oct. Bn. briefly met 3^{e}BCCP and Co. Loth at Hanoi/Gia Lam before cloud break allowed those units to take off for That Khe, and annihilation. 2^{e}BEP only prevented following them by resumed cloud cover, solid for two days. When extent RC 4 disaster realised, these two cos. retained Hanoi—only remaining airborne reserves in Tonkin. 24 Oct., remainder bn. arrived Hanoi.

3 Feb. 1951 Under new CO, Ch. d'Esc. Raffali, successful actions near Ke Sat, central Delta.

Apr. 2^{e}CIPLE assigned to bn.; initial misgivings over local recruitment largely overcome by good performance in May at Phat Diem during VM Day River attacks.

Summer Continual security ops. in Delta.

4 Oct. Bn. jumped Gia Loi, Thai Highlands, supporting 8^{e}BPC in attempt cut Khau Vac track, supply route VM 312 Div. then advancing S on important post Nghia Lo. Severe fighting in punishing mountain jungle, Ban Van-Tan Kouen area: monsoon rains, bad radio reception, no food, heavy combat with VM Regt. 209; 53 casualties.

16 Dec. 1951–12 Jan. 1952 Heavily engaged on Ba Vi massif during 'Black River' battles.

13 Jan.–24 Feb. Bn. held first Hill 202, then Suc Sich, posts on RC 6 during last stage Hoa Binh battles. 24 Feb., pulled out down RC 6 during leapfrog French withdrawal from untenable Hoa Binh enclave; heavily ambushed.

Mar.–Aug. Continual security ops. in Delta.

1 Sept. Two days before end of two-year tour, much-liked CO Lt. Col. Raffali mortally wounded by sniper during routine op.

9 Nov. Op. 'Marion'—see under 1^{er}BEP.

19 Nov. Bn. flown into Na San.

23 Nov.–2 Dec. Defence of Na San—see under 1^{er}BEP. Bn. held PAs 4, 21bis, 22bis.

30 Dec. Hard fighting Co Noi during French probes after withdrawing VM.

18 Apr.–early July 1953 Bn. operated Plain of Jars, Laos.

16 July Recalled Hanoi under cover 14 July parade, bn. jumped with 6^{e}BPC, 8^{e}GCP in airborne phase Op. 'Hirondelle'—Gen. Gilles's brilliantly successful raid on VM dumps around Lang Son, coordinated with thrust up RC 4 by Groupe Mobile 5. 2^{e}BEP jumped Loc Binh, half-way to Lang Son; repaired and held river crossing for later withdrawal other para bns. falling back after massive destruction VM materiel. Spearhead paras arrived Loc Binh by forced march early 18 July, passed through BEP, who formed rearguard further retreat Dinh Lap, where RV'd with road column; fatalities from heatstroke on march. Trucked Dinh Lap-Tien Yen night 19/20 July, force successfully extracted by sea.

Aug.–Sept. Continual security ops. in Delta; many casualties mines, booby-traps, snipers.

Jan. 1954 Ops. Thakhek, Laos; Saigon, Nha Trang.

20 Jan. Op. 'Aréthuse'—amphibious landing, Tuy Hoa.

30 Jan. Returned Nha Trang; then to Pleiku, seven weeks' punishing ops. in Central Highlands.

18 Mar. Moved urgently Hanoi; awaited order to jump into Dien Bien Phu.

Dien Bien Phu

Shortage of space naturally prevents more than the briefest note on this long and complex battle; readers are recommended to acquire Bernard Fall's classic *Hell in a Very Small Place* for the best English language account.

In late 1953 the French C-in-C, Gen. Navarre, decided to plant a fortified camp, entirely dependent on an air link, in the valley of Dien Bien Phu in the Thai Highlands 220 miles from the Delta. It was to be an anchor for French-led commando groups operating in the Laotian border country; and a base for offensive sorties by powerful units supported by artillery and aircraft based inside the enclave. Although it quite soon became apparent that offensive operations were impossible, due to heavy enemy presence in the surrounding hills, the camp was maintained. It was now the command view that it would be the anvil on which artillery and air power could smash Giap's regular divisions. The implications of the camp's new rôle were not thought through. The size of the Viet artillery; Giap's ability to move guns and ammunition across hundreds of kilometres, and to install them round the valley; and that artillery's chances of survival under French air and counter-battery attack, were all chronically underestimated. The valley was far too large for the garrison provided to hold the commanding hills all around. Engineer stores were hopelessly inadequate to protect the central area and the loose ring of mutually supporting perimeter strongpoints against artillery. And finally, the French did not anticipate that Giap now had a regular flak regiment.

The valley was taken without difficulty on 20–22 November 1953 by six para battalions, including the 1ᵉʳBEP and its 1ᵉʳᵉCEPML. The Legion unit was one of two para battalions which remained at Dien Bien Phu when the rest withdrew, being replaced by ten air-lifted infantry battalions: four Legion, three Algerian, one Moroccan, and two Thai. The camp received 24 × 105 mm and 4 × 155 mm guns; ten M24 tanks; and six Bearcat fighter-bombers. The total garrison was more than 10,000 men, but several thousand of these were not combat infantry.

Giap accepted the challenge; and between November 1953 and early March 1954 he successfully gathered in the surrounding hills some 49,500 men of the 304, 308, 312 and 316 Divs. and the 351 Heavy Div., supported by more than 200 guns of 75 mm and 105 mm calibre, and several score heavy mortars and RCLs. He also installed his Chinese-trained 37 mm anti-aircraft regiment.

Sorties by the garrison in December quickly demonstrated that even three para battalions operating with tank, artillery and air support could not get further than a few miles out of the valley without serious fighting. Soon, patrols on the valley floor itself were involved in daily firefights. On 31 December a desultory shelling began.

On 13 March 1954 a massive and accurately directed artillery barrage fell on the camp, opening the battle proper. Within three days two vital northern strongpoint systems had fallen, with the loss of two strong battalions. The air force proved unable even to find the brilliantly dispersed and camouflaged Viet artillery; the camp's guns proved unable to silence it, and indeed, began to take heavy punishment themselves. The flak soon reached Second World War intensity; and the airstrip was shelled with murderous accuracy. The last 'casevac' Dakota managed to get off on 27 March. From then on all supplies and reinforcements had to be para-dropped, into a constantly shrinking perimeter, usually at night, by aircraft running the gauntlet of heavy flak. The garrison's misery, and the air force's

cont. on page 16

DIEN BIEN PHU

(*Left, top*) November 1953—120 mm tube of 1^{ere}CEPML during early stages of Operation 'Castor'. All native houses were later torn down for fortification materials.

(*Left, centre*) Col. de Castries outside his command bunker. This gives some idea of the style of fortification: sandbags, earth banks, and thin wiring, with tents pitched over some open dugouts. The tall radio masts made an easy aiming point for enemy bombardment of command posts.

(*Left, below*) Dien Bien Phu in November 1953, looking south and slightly east. The parked C-47s are on what became the southern end of the enlarged airstrip. At top left is the terraced hill which became 'Eliane 2'. By the opening of the siege proper all these trees had been cut down for timber, and much of the brush cleared for fields of fire. By April the valley floor was a moonscape of mud and shell holes. (ECPA)

(*Right, top*) Imaginary layout of a company sector on a battalion's hill-feature strongpoint. Few positions actually had two complete belts of trenches. Not enough wire was available for complex entanglements—most strongpoints simply had repeated lengths of 'cattle fencing' on timber pickets. Overhead protection was poor: few bunkers had the three feet of packed earth on solid timber roofing necessary to stop field artillery, and most were improvised from light timber and sandbags. It was later calculated that to fly in sufficient engineer stores would have tied up the whole French airlift capacity in Indochina for five months non-stop. Key: (B) Co. 'blockhouse' bunker with MGs; (C) Co. HQ bunker; (P) Ptn. HQ bunkers; (HQ) Bn. HQ bunker; (A) Auxiliary bn.HQ, and officers' mess; (MO) Medical bunker; (S) Signals bunker; (M) Bn. mortars. (Christa Hook)

(*Right, centre left*) Clearing 'Eliane 2', December 1953. The trees have been cut; and the brick-built 'Governor's House' is being demolished. Its strong cellar was later used as a CP on this vital position, where the 1^{er}BEP saw murderous fighting. (ECPA)

(*Right, centre right*) Paras, of a Colonial battalion, pile their packs on the parapet of a rudimentary trench. The two foreground figures wear a British windproof smock, and an American camouflage jacket.

(*Right, below*) Part of 'Eliane' after clearance and digging in; the figures in the right background give scale. We are looking east, at the wooded hills held by the Viet Minh. (ECPA)

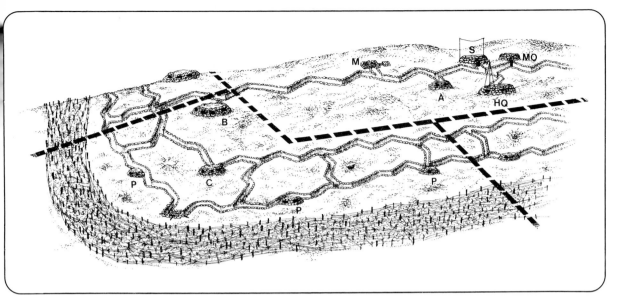

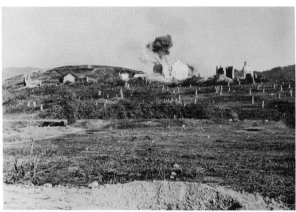

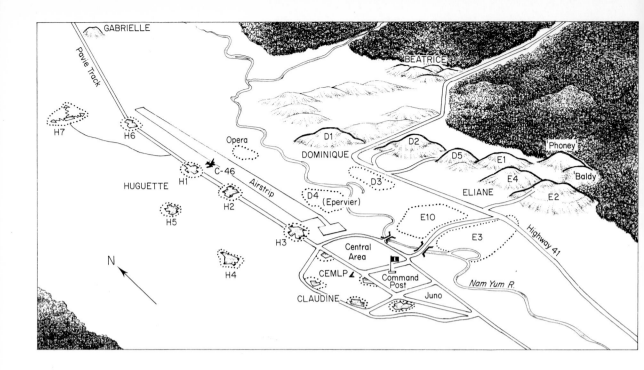

An oblique and much simplified view of the main Dien Bien Phu position, looking from south-west to north-east. The north-eastern strongpoint 'Beatrice' was overrun on the night of 13 March; the northern strongpoint 'Gabrielle' fell the following night. This allowed a fatal tightening of the Viet siege ring and, particularly, brought enemy artillery and AA positions much closer to the airstrip. 'Dominique 1' fell on 30/31 March and was never retaken, seriously compromising the rest of the camp. At the south-east corner the two hills 'Phoney' and 'Baldy' ('Mont Fictif' and 'Mont Chauve') were in enemy hands throughout the siege. 'Baldy' was connected by a saddle to 'Champs Elysées', the south-east slope of 'Eliane 2', and was the jumping-off point for many mass attacks during the 'battle of the five hills' in early April, when the 1ᵉʳBEP fought on 'Eliane'.

An attempted counter-attack towards 'Huguette 1', cut off by enemy approach trenches, cost the 2ᵉBEP dear on 23 April. The wrecked C-46 transport was used as an MG nest by the Viets; even a few feet of elevation gave them an important advantage over the paras, who were caught in the open out on the airstrip. (Christa Hook)

problems, increased on 29 March when the monsoon rains began. The weather made air strikes and supply drops even more uncertain from that date on; and under the impact of rain and constant shellfire, the inadequate trenches and bunkers began to melt into a sea of red mud.

The character of the infantry fighting recalls the worst episodes of the battle of Verdun. Under constant pounding from Viet artillery, the dwindling combat units at the disposal of Lt. Col. Langlais and Maj. Bigeard held, lost, retook, and lost again a series of vital hillocks and holes in the mud, in the face of 'human wave' night attacks. The enemy pushed saps and trenches in from all sides, first

approaching, then isolating, then strangling one strongpoint after another. Epics of courage were recorded as the paratroopers of the counter-attack reserve, reeling from fatigue and short of every necessity, carried out sacrificial attacks to retake lost positions, or to hack a way out for surrounded companies. Several fresh para battalions, and hundreds of untrained volunteers of other branches, jumped into this small hell on earth during the siege, right up to the eve of the final collapse.

Some of the more important dates from the viewpoint of the Legion paras were as follows:

21 Nov. 1953 1ᵉʳBEP, 1ᵉʳᵉCEPML jumped into Dien Bien Phu on second day of Op. 'Castor'.

10–15 Dec. 1ᵉʳBEP lost 52 men during three-bn. sortie north from valley.

23–28 Dec. 1ᵉʳBEP took part in two-bn. sortie south into Laos as high command publicity stunt.

12–16 Feb. and 5 Mar. 1954 1ᵉʳBEP took part in unsuccessful sorties to clear enemy from hills around northern strongpoint 'Beatrice'.

13 Mar. Opening barrage of battle proper; 'Beatrice' overrun in night attack.

14 Mar. Strongpoint 'Gabrielle' overrun.

15 Mar. Two cos.1ᵉʳBEP took part unsuccessful counter-attack towards 'Gabrielle'.

25 Mar. 1ᵉʳBEP in action on eastern 'Eliane' strongpoints.

26–27 Mar. 1^{er}BEP attacked enemy penetrations around 'Huguette 6' and '7', isolated at far end of airstrip.

30 Mar. 1st Co., 1^{er}BEP took over positions on 'Eliane 2'. Start of mass enemy attacks on the 'five hills' of 'Dominique' and 'Eliane'. By midnight, VM established on 'Champs Elysées', lower part of 'Eliane 2'.

1 Apr. 1^{er}BEP took part in successful counter-attack on 'Champs Elysées', but enemy renewed attacks that afternoon.

2–4 Apr. Repeated, heavy attacks on 'Eliane 2' finally slackened after 107 hrs of virtually non-stop fighting.

9/10 Apr. First of three lifts to drop 2^eBEP into camp, completed 12 Apr.; one co. sent at once to 'Dominique 3', two to 'Dominique 4'. In first 12 hrs on ground, bn. lost 29 men simply to random shellfire.

10/11 Apr. Two cos. 1^{er}BEP took part in recapture of lost strongpoint 'Eliane 1'—eyewitnesses confirm that paras went into attack singing. Co. from 2^eBEP moved from 'Dominique 3' to consolidate on 'Eliane 1'.

17/18 Apr. 1^{er}BEP's attempt to break through to cut-off garrison of strongpoint 'Huguette 6' failed.

23 Apr. 2^eBEP's attempt to retake lost strongpoint 'Huguette 1' failed; bn. pinned down on airstrip, suffered disastrous casualties—c. 150 out of 380 men. (Wounded Lt. Garin seen to shoot himself dead to prevent his men risking themselves to bring him into cover.)

25 Apr. Survivors of both BEPs grouped in composite 'Bataillon de Marche' under Ch. de Bn. Guiraud, CO 1^{er}BEP; assigned defence of southern 'Huguette' strongpoints. Relative lull in mass infantry attacks until

30 Apr./1 May VM 308 Div. took 'Huguette 5'; immediately driven out by BEP, at cost of 88 French casualties.

2 May Counter-attack by BEP on lost strongpoint 'Huguette 4' failed.

6/7 May Final VM assault. Of c. 160 effectives remaining in composite BEP, holding 'Huguette 2' and '3', some 80–100 sent right across camp in response to appeal for support from collapsing 'Eliane' strongpoints. Many killed by shellfire on way; remainder overrun with other para units on 'Eliane 4' and '10'.

7 May, 1730 hrs Gen. de Castries, camp CO, ordered ceasefire after fall of whole eastern half of camp.

Exact figures are not known: one source states that of 635 men who jumped with the 1^{er}BEP, 575 were posted killed or missing; it is likely that only about 50 men remained to the 2^eBEP by 7 May; and 32 men of the 1^{ere}CEPML went into captivity. The numbers who survived the 'death march' and the Viet Minh camps were considerably smaller.

Within three months France had agreed to a general ceasefire, and to the partition of Vietnam.

*   *   *

Neither BEP was disbanded after this disaster; rear base personnel, and returning wounded evacuated before 27 March kept both units officially alive until replacements arrived from North Africa. On 25 May 1954 the 3^eBEP arrived at Haiphong, and collectively took over the identity of the 2^eBEP. Based at Hanoi/Cat Bi until 21 July, the unit then flew south to Cochinchina, where it remained until embarkation for Algeria on 1 November 1955. The 1^{er}BEP, also re-formed from replacements and returning veterans, moved from Haiphong to Hue on 20 August 1954. Under the command of Capt. Jeanpierre, the battalion sailed for Algeria on 8 February 1955.

The Algerian War 1954–62

On 1 September 1955 the 1er and 3^eBEP were redesignated as 'Régiments Étrangers de Parachutistes'; but on 1 December the 2^eBEP was also raised to regimental establishment, and the 3e (which had seen some action in July–November) was disbanded, its effectives being split between the two senior units. The accompanying Table 2 shows the new regimental establishment. Both REPs earned a high reputation as units of the 'general reserve' throughout the Algerian War; but this brief summary does not allow a detailed record of their operations—which would, in any case, be wearisomely repetitive. A few paragraphs on the character of the fighting, and on one or two important and typical operations, may help to put the légionnaire-para's war in perspective.

Despite superficial similarities, the Algerian War was quite unlike the Indochina War in essentials. The Muslim nationalist 'Army of National Liberation' (ALN) followed the Maoist revolutionary war blueprint; they began their campaign with low-intensity guerrilla attacks against widely dispersed 'soft targets', and began to build up conventional units in camps across the Tunisian and Moroccan borders. Their initially patchy support among the native population hardened as time passed. The ALN provoked excesses by the European settler community and the French Army, and terrorised waverers by selective atrocity; and large-scale

14 July 1954: the 2ᵉBEP, reborn by the wholesale redesignation of the newly arrived 3ᵉBEP, takes part in Hanoi's Bastille Day parade. These NCOs all wear 47/52 airborne camouflage fatigues, with para wings, unit badges, medals and citation lanyards—the latter apparently worn without the 'olive': see commentaries Plates E1, K9. (ECPA)

forced 'resettlement' of villagers from combat zones caused great misery. The rebels had the run of a huge, thinly populated country, much of it a mountainous wilderness. But for all this, the French achieved an undeniable military victory; Algerian independence was won, in 1962, by political rather than military means.

The Muslim population was far from unified. French Intelligence penetrated the ALN, and exploited regional and personal enmities to foment bloody fighting between different factions. France also launched an imaginative 'hearts and minds' programme; in 1959 nearly 30,000 *harkis*—Muslim auxiliaries, some of them 'turned' ALN fighters— were serving with the French Army. (This was about twice the active strength of the ALN inside Algeria at that time.) By late 1959 ALN morale had been hit so hard by French Intelligence coups, military defeats, and internal feuding that there were large-scale defections; and one group of senior commanders flew to Paris for secret ceasefire talks.

Far from building their guerrilla bands into large, conventional units for campaigns of manoeuvre, the ALN were forced by 1960 to disperse much of their remaining strength into gangs of ten or 20 fugitives.

Apart from areas of cork, scrub oak and cedar forest in the coastal hills, the wild interior of the country has little continuous overhead cover. An army can certainly hide in the gullies, boulder fields and scrub-covered slopes of the Aurès and Kabylie mountains—but only until it moves. A French air force far stronger than that which fought in Vietnam recorded many successes in interdicting daylight cross-country movement by ALN units.

The 'main force' units raised in Tunisian and Moroccan camps were limited by the fact that Arab support was more vocal than practical. Most weaponry had to be bought on the open market; and these units never approached the standard achieved by Giap's *Chuc Luc*. Most important of all, the French succeeded in denying them access to the battlefield.

During the first two years after the outbreak of war on 1 November 1954 the ALN were able to increase their strength and multiply their guerrilla attacks with some success. The large French army in Algeria—some 200,000 even by early 1956— were mostly static 'sector troops', guarding the 1,200,000 white settlers and the loyal natives who were the ALN's prime targets for terrorism. Most 'sector troops' were conscripts (the legal fiction that Algeria was 'part of France' allowed their deployment), but initially even high quality regular units of the Legion and the airborne corps were split up into 'penny packets'. Counter-insurgency successes were uneven, depending largely upon local factors. However, between summer 1957 and

An officers' O-group of the re-formed 1^{er}BEP near Hue, summer 1954. All wear the 47/51 or 47/52 fatigues and 1950 webbing. Details to note: the left man has a leather holster tucked behind the hip tightening tab of his smock, a common practice; the second from left has a first aid pouch tied to his upper arm; the third from left has a field dressing tied to an old Mauser bayonet worn as a fighting knife. (EPCA)

TABLE 2

**1er and 2e Régiments Étrangers de Parachutistes
Algeria, 1955–61**

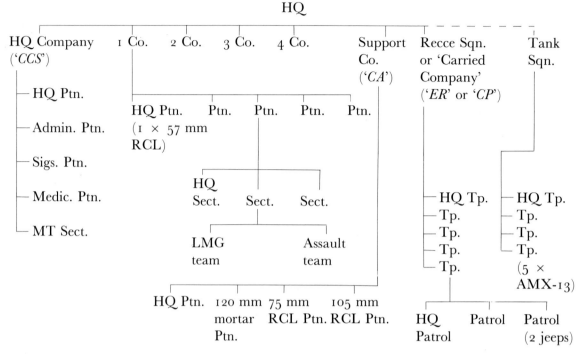

Notes:

The tank sqn. was unique to 1^{er}REP; transferred from 2^{e}REC for Suez landings, Nov. 1956, it was retained; crews fought mostly as infantry. The Support and Recce Cos. drove MG-armed jeeps, and Dodge 4×4 and 6×6 trucks.

Total regimental strength, c.1,300, with variations. At one point, e.g., 2^{e}REP's CCS reached strength of 344.

spring 1960 the whole rhythm of the war changed in France's favour.

The 'Morice Line'—a broad barrier of electrified wire, mines and radar sensors—was completed in September 1957, stretching along the Tunisian border from the sea down to the empty sands of the Saharan fringe, where nothing larger than a scorpion could move unseen from the air. A similar barrier faced the smaller ALN camps across the Moroccan frontier. French mobile units were deployed all along the Algerian side of the borders. In the winter of 1957–58 there were a series of heavy engagements, as large ALN supply and reinforcement units attempted to break through by night. Nearly all were stopped dead at the wire, or hunted

down within 48 hours of crossing, before they could melt away into the mountains. The 'battle of the frontiers' lasted some seven months, and the two REPs played an active part in it. By the time the ALN crossing attempts petered out, they had cost the rebels some 6,000 dead.

The French then settled down to a systematic clearance of the major refuges in-country: the rugged crests and gorges of the Aurès, the Nementchas, the Kabylie highlands, the Hodna, the Collo Peninsula, eastern Constantine, and the Ouarsenis hills. The climax came in 1959–60, with Gen. Challe's great operations 'Couronne', 'Courroie', 'Etincelles', 'Jumelles', 'Pierres Précieuses' and 'Trident'. By spring 1960 several thousand

ALN fighters had been killed; and—unusually, and significantly—several hundred had chosen to surrender.

These operations, in which the REPs were heavily involved, depended upon the speed, flexibility, and co-ordination of large ground and air forces (and occasionally, even coastal landings by Marines). Acting on Intelligence reports, 'hunting commandos' of French and *harkis* would probe a target sector until they made contact; and would maintain that contact at all costs, while calling in the motorised and para units of the 'general reserve'. These mobile units might total as many as 25,000 for a major operation, apart from the static 'sector troops' already in place. All elements would be co-ordinated by a single HQ.

By the skilful use of map, radio, spotter 'plane and helicopter—and by a great deal of night driving followed by hard marching over appalling terrain from the nearest 'truckable' roads—the paras would weave a tight net all around the location. Helicopters would be used to block the most immediately vulnerable escape routes, and to lift radio scout squads on to commanding heights; they would then remain available to the operation commander for tactical use in the coming battle. The experienced para commanders acquired impressive skill in lifting sub-units around the battlefield under fire with pin-point accuracy, reacting to enemy movements reported by spotter aircraft or scouts on the crests.

(Although jump-trained, French paras made very few combat jumps in Algeria; they were employed as élite light infantry and 'air cavalry'. Even in 1954, the French in Indochina had only about half a dozen light 'casevac' helicopters; by the end of the decade they had enough airlift capacity in Algeria to carry two battalions simultaneously.)

When the net was complete, the paras would start their search-and-destroy sweeps. These might last several days and nights, and were extremely punishing—the Algerian hinterland is inhospitable at all seasons, and very little of it is horizontal. The paras would search every slope, every gully, every boulder and every cave (a particularly nightmare task) until they took fire from the elusive ALN *katiba*. They would then maintain contact until the enemy unit was finally trapped; and would wipe it out by co-ordinated infantry, air and artillery

In November 1956 the 10ᵉ Division Parachutiste was withdrawn from Algeria to take part in Operation 'Musketeer', the Anglo-French landings at Suez. The paras impressed British observers by their professionalism; and like the rest of the task force, they were intensely frustrated by the ceasefire imposed by US and UNO pressure. Returning to Algeria in December, the 1ᵉʳREP retained the AMX-13 tank squadron from the 2ᵉREC which had been attached for their amphibious landing at Port Said. This photo shows an FM 24/29 team of the 1ᵉʳREP near the Canal. (ECPA)

attacks. The operation would continue, wherever it led and however long it took, until every enemy unit in the sector had been found, destroyed, or broken up into tiny fugitive bands. When the sector was judged controllable by the 'sector' troops, supported by 'hearts and minds' teams, the 'Challe Steamroller' would move on to the next sector. The troops lived under canvas, in all weathers, for weeks at a time.

If this sounds easy, the reader is invited to imagine the task of the infantryman, sweeping slowly across very broken terrain, searching for a desperate enemy concealed among the rocks and scrub. The ALN units were often in company strength, well armed with Second World War small arms and supporting weapons, and determined to sell their lives dearly. Total ALN casualties were, typically, ten times those of the French; but this statistic was of little consolation to the individual para squad who lost half their mates in some ugly, point-blank firefight which flared up without warning in thick cover. An operation might easily cost a unit half a dozen dead and 20 wounded, usually concentrated in a single company; and junior officers and NCOs suffered dispropor-

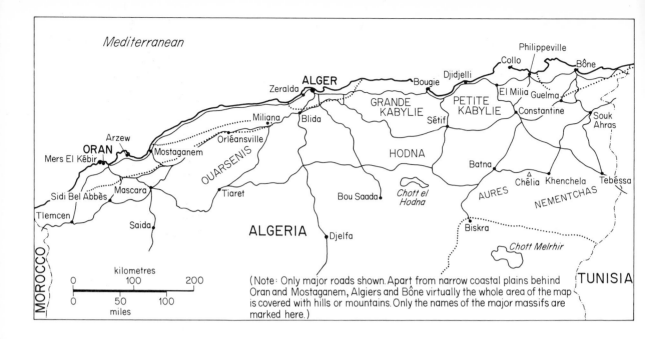

(Note: Only major roads shown. Apart from narrow coastal plains behind Oran and Mostaganem, Algiers and Bône virtually the whole area of the map is covered with hills or mountains. Only the names of the major massifs are marked here.)

Simplified map of northern Algeria. The 1ᵉʳREP had its rear base at Zeralda, the 2ᵉREP at Philippeville, but both units served for months at a time at other, forward bases. The 1ᵉʳREP saw action all over the country; the 2ᵉ, mainly in the east of Constantine province. They served successive tours at Guelma during the 'battle of the frontiers' in 1957–58.

tionately. The REPs spent most of their war out in the *djebels* on this kind of operation, and casualties mounted steadily. The war as a whole cost the 1ᵉʳREP 775 casualties; the 2ᵉREP, 741. (But the 2ᵉ recorded more than 3,650 ALN killed and 538 captured, which puts these losses into perspective.)

Several para units achieved remarkable successes in these operations; with sister regiments like Bigeard's 3rd Colonial Paras, the REPs were hardly in a one-horse race. But the 'green berets' became greatly respected even among this demanding peer-group; and the morale, suppleness and 'punch' of, say, the 1ᵉʳREP during Lt. Col. Jeanpierre's period of command in 1957–58 became legendary. This hard-driving commander, who was killed when his Alouette helicopter was brought down by ground fire in May 1958, made the regiment very much his own instrument. He organised it and trained it in such a way that any number of companies could be placed, at short notice, under either of two tactical HQs; the two combat groups became so expert that

Fine study of 1ᵉʳREP sentry outside an HQ tent at Suez, 1956; he wears an early smock, the later trousers, and double-buckle boots, and carries an M1A1 carbine. (ECPA)

22

they could even swap companies in the middle of a firefight without losing cohesion. All ranks, including the motorised companies and the tank squadron were expected to meet the same exacting standards of physical endurance and infantry skills.

Although few privates or lieutenants had fought in Indochina, the captains, majors and senior NCOs were nearly all scarred and decorated veterans. Company and combat group commanders like Brandon, Domigo, Caillaud, Allaire, Carvalho, 'Lulu' Martin and Cabiro made the Legion paras richer in battle-proven leadership than almost any contemporary Western unit. Spurred on by the memory of sacrifice and defeat, and by determination to 'win this one' at whatever cost, the paras were extremely dangerous troops.

A typical operation on 26–27 April 1958 found

Légionnaire Walter Krainer poses for a studio portrait soon after joining the 2ᵉREP in Algeria, early 1960. His regimental citation lanyard—*fourragère*—is of the 'fancy' type with an extra loop or '*tour-de-bras*'. (Richard Leach)

the 2ᵉREP in action at Beni Sbihi south-east of El Milia. Helicoptered in among 3,000-ft wooded crests while other units moved into blocking positions round a reported ALN *katiba*, the 2ᵉREP's Lt. Col. Lefort made a lightning change of direction as a result of a suspicious contact. His 'nose' for ground and the enemy proved acute: taking the gamble of tearing up his original orders, he led his regiment to a victory which cost the ALN 199 dead, for REP losses of just five killed and 26 wounded.

As the 'Challe Plan' unfolded, the 'green berets' profited by the chance to practise their battle skills

Légionnaire Lambrecht, 1ᵉʳREP, photographed in Algeria, 1955. Sweaters, scarves and toques were necessary during winter operations in the harsh, mountainous interior, where freezing rain and heavy snow were not unusual.

Paras line up for a lift in Boeing-Vertol H-21 Shawnee helicopters—'flying bananas'. The first of these arrived in Algeria at the end of 1957; each could carry 20 men—two infantry sections. Although there were occasional battalion assaults by helicopter, it was more usual for only one or two companies at a time to be inserted by air. By 1960 the helicopters were usually grouped in units of four H-21s, with an Alouette command ship and a Sikorsky H-34 (British equivalent, Westland Wessex) fitted as a 'Pirate' gunship with a 20 mm cannon and two .50cal. MGs.

after a quieter period. Between July 1959 and January 1960 the different phases of Operation 'Jumelles' allowed the 1^{er}REP to accumulate a record of 513 enemy killed, 222 captured, and 428 weapons recovered. The dispersal of the surviving enemy units meant that major contacts became rare during 1960, though the backbreaking operations continued, and so did a steady haemorrhage of casualties. It was allegedly the 'last *katiba* in the Aurès' which the 2^{e}REP trapped at Chélia on 2 December 1960 during an operation set up by the much-liked second in command, Cdt. Cabiro. On this anniversary of Austerlitz he led the regiment in a wild, scrambling fight up and down the thickly wooded slopes, finally achieving 54 enemy killed for own losses of ten dead and 19 wounded—a most unusual encounter for that stage of the war.

*　　*　　*

Even in such a brief account as this, there is no point in evading the two most controversial episodes in the Legion paras' war: the 'battle of Algiers' and the 'generals' *putsch*'. To explore adequately the arguments raised by each would take several thick books; here, a few paragraphs must suffice.

The Battle of Algiers

In winter 1956–57 the ALN, which had established a strong network of perhaps 1,200 activists within the Algiers city *casbah*, carried out a major urban terrorist campaign. Sallying out in disguise from the maze of interconnecting houses, alleys, courtyards, rooftops and stairs where 100,000 Muslims lived in just one square kilometre, the girl bombers killed and maimed scores of European men, women and children. The police proved incapable of stopping the slaughter, and public feeling ran high. In mid-January Gen. Massu's 10th Parachute Division was ordered into the city[1]. Given wide powers under a state of virtual martial law, the paras quickly

[1] 1^{er}REP, 1^{er}RCP, 2^e and 3^{e}RPC.

24

unravelled much of the ALN network, and inflicted severe losses on the terrorist infrastructure.

By the time they returned to the *djebels* in mid-April 1957 they seemed to have drawn the ALN's teeth; but a renewed bombing campaign in June brought them back on 1 September to finish the job for ever. The two surviving ALN leaders were both trapped by the 1^{er}REP: Lt. Col. Jeanpierre was wounded leading the group which captured the remarkable Yacef Saadi on 24 September, and Ali la Pointe was blown up in his hiding place on 8 October. Even the ALN admitted that their organisation in Algiers was, for all practical purposes, dead. And even the paras' most convinced critics admitted that the means they employed were probably the only ones which would have worked under those circumstances.

Nobody, from Massu down, has ever attempted to evade the fact that the rapid unravelling of the ALN organisation was achieved partly by the widespread use of torture. Massu himself submitted to torture by electrode before authorising its use on suspects. He argued that, having laid upon the paras the task of stopping the bombers without delay and by military force, the French government and people could hardly disassociate themselves from the only possible means of carrying out that mission. The first 24 hours of interrogation are vital, if information is to be extracted before the torn threads of the terrorist organisation can be sealed off. It is also undeniable that the ALN was guilty of the most atrocious tortures, both as punishment and as warning. Whatever one's views on the moral issue, two facts remain: innocent lives were saved by the destruction of the ALN's bombing network; and the degradation of torture rubbed off, in some cases, on the torturers. In the long term, this episode damaged French morale.

The Putsch

Despite her military success, France's political response to the war was confused. A series of weak and short-lived governments became a scandal; and in May 1958, with the connivance of many officers, the Fourth Republic fell. Gen. de Gaulle was summoned to power as president, in the expectation that he would prosecute the war firmly.

As the next three years passed, it slowly emerged that de Gaulle was accepting the historical

The menacing elegance of the '*tenue léopard*', symbolic of the Algerian period; and the two essential tools of the 'Challe Plan'—the map and the radio. A Legion para officer with his HQ group photographed on the *djebels* in 1959.

inevitability of Algerian self-determination. Hopes for some compromise solution, which would protect French and settler interests in a semi-independent Algeria, were destroyed by the stubbornness of militants in both settler and Muslim communities.

25

The Army, which had fought so hard to win a war for France, and which had not been afraid to risk its good name as well as its blood, now faced the certainty that it had all been for nothing. Algerian independence under these conditions would ruin a million and a quarter settlers, who would lose everything their families had built over 130 years. It would orphan the Foreign Legion, which existed to garrison Algeria. And it would doom the loyal *harkis*, who had been promised that France would always stand by them[1].

Men of transparent decency and integrity, such as Gen. Challe, and Cdt. Denoix de Saint-Marc of the 1^{er}REP, were faced with an agonising moral dilemma. It could be argued that de Gaulle himself had come to power through chosing to disobey his

Operation successful: the 1^{er}REP assemble ALN small arms after a clash with the rebels. The left hand para (who wears the 10^{e}DP patch on his shoulder, which was not common during combat operations) carries a Czech-made Mauser, a shotgun, and a Garand M1 with its butt covered in sacking. The central background figure has a yellow and black air recognition panel tied to his pack: a necessary precaution during operations which often involved very close tac-air support in scrub-covered terrain—'own casualties' from rocket and cannon fire were not unknown.

[1]After French withdrawal, it is estimated that some 100,000 were butchered, often with medieval cruelty. Only about 15,000 found safety in France.

legally elected and appointed superiors in June 1940. (Any British readers tempted to adopt a superior attitude toward the 'politicised' French Army should reflect upon French history since 1870; and thank their God in silence.)

On 22 April 1961 the 1ᵉʳREP, with the 14ᵉ and 18ᵉRCP, led an almost bloodless coup in Algiers under the command of four retired generals who included two former commanders-in-chief in Algeria—Salan and Challe. It is hard to understand exactly what they intended to do with their mutiny had it succeeded: they were motivated by a determination not to see the country handed over to the ALN, but they did not really think through how this was to be achieved. Officers of the 2ᵉREP, who flirted briefly with the *putsch* during its later stages, detected an air of unreality and indecision. The whole adventure collapsed in days, without further bloodshed. The bulk of the armed forces refused to rally to the mutineers, whose fate was sealed by de Gaulle's masterly radio broadcasts to the troops. On 30 April the 1ᵉʳREP was disbanded, without resistance. The officers, together with many from other units, were arrested, tried, dismissed the service, and in many cases imprisoned until an amnesty in 1968. The men were dispersed throughout the Foreign Legion.

The men of April 1961 were not all as admirable as Challe and Saint-Marc; some later joined the murderous gangsters of the OAS; but to dismiss the best of them as unscrupulous political adventurers is simply impertinent. Their fate is a reminder of the dreadful consequences which can follow, even in a modern democracy, when a government creates a military élite and then uses it as if it were an unthinking machine. The final chapter in the story of the regiment of RC 4 and Dien Bien Phu can only leave us with a tragic sense of waste.

Para-Commandos

The *putsch* more or less destroyed the French bargaining position at the Evian talks, and marked the end of major operations. The official ceasefire came on 19 March 1962; but the agreement allowed certain French units to stay in Algerian base areas for a transitional period. After a last miserable year of hard marching, bitter little clashes, and low

An officer and his 'radio'—in the French Army, this refers to the operator—wearing the olive 'Satin 300' fatigues which the 2ᵉREP have worn since the beginning of the 1970s. Apart from the vertical chest pocket zips the most obvious recognition feature is the very tightly tailored cut favoured by the paras; the leg cargo pockets have no gussets, and are worn almost flat to the thigh. (ECPA)

morale, the 2ᵉREP was moved in March 1962 to Telergma, and in September to the Mers-el-Kébir base area. Here Lt. Col. Chenel[1] kept them busy building a new camp in an unpromising coastal swamp at Bou Sfer.

After this grim period of hard, apparently pointless labour in an atmosphere of distrust, when the life of the regiment seemed to hang by a thread, the arrival of Lt. Col. Caillaud in June 1963 brought an electrifying improvement. His plan was to transform the 2ᵉREP from a conventional parachute infantry unit into a uniquely skilled

---

[1]The same officer who had started his jump training at Kun Ming in 1945—but who had only found time to complete it in 1961! He had been the first 'Legion para' of all: and in 1961–63 his calm, protective attitude may well have saved the Legion paras from oblivion.

The 2ᵉREP's *Equipe de Chuteurs Operationnels* **prepare for a jump in Corsica, 1970. This 'pathfinder' sub-unit was introduced as part of the specialisation programme instituted by Lt. Col. Caillaud in 1963–65. They are highly trained in advanced parachuting techniques, including free-fall HALO jumps and long lateral approaches with ram-air canopies. This photo shows the MAS 49/56 rifle thrust behind the ventral pack, and the rucksack strapped high underneath it. The pathfinders, whose mission is primarily deep reconnaissance behind enemy lines, have had various designations over the past 20 years. Currently a platoon within the regiment's Recce and Support Co., they are designated** *Commando de Renseignements et d'Action dans la Profondeur* **or CRAP. It would be extremely hazardous for any English-speaker to attempt to point out the humour of this acronym to members of the unit.**

para-commando. Officers and men would master an unprecedented range of military techniques— amphibious operations, mountain warfare, anti-tank combat, mining and sabotage, advanced parachuting skills, deep reconnaissance—and would fit themselves to take on any task, at any time, anywhere in the world. This inspiring vision rekindled the regiment's enthusiasm; it gave them a worthwhile goal, and the prospect of élite status once again. In two years Caillaud achieved extraordinary results. Recruitment would become highly selective, as the new reputation of the unit attracted great interest. Re-enlistment rates soared, in all ranks.

In June 1967 the whole regiment was installed at

its new base, Camp Raffali at Calvi, Corsica; advanced parties had been preparing for the move for two years past. The 2ᵉREP now joined the order of battle of the 11th Division, as part of France's rapid overseas intervention forces—a rôle for which its special and always developing skills fitted it perfectly. Given France's continuing defence links with her former colonies in sub-Saharan Africa, the chances of active service—unimaginable five years before—were distinctly improved.

Tchad

The chance came in April 1969, when a tactical HQ ('EMT 1') and the regiment's 1st and 2nd Cos. were sent to join French elements supporting the then-President Tombalbaye of Tchad against two separate rebel groups. One was active in the east, along the Sudanese border; the other, basically a Tubu tribal movement, in the Saharan emptiness of the northern region. The sub-units spread out from a base at Mongo, patrolling, and dealing with any bands they encountered. In September the regiment's 'EMT 2' arrived, with the 3rd Co. and the CEA (Recce and Support Co.); and a motorised company (CMLE) drawn from other Legion units.

With the southern regions more or less cleared, attention turned mainly to the aching void of sand, baked rock and scrub below the Libyan border in the north. 'Nomadising' all over the vast country by truck, helicopter, and even on horseback, the légionnaires learned about things which had been all too familiar to the '*grands ançiens*' of the '*Légion du Papa*': the sun that kills, the choking dust, the horizon dancing in the heat, and the eeriness of a landscape that was never meant for human beings. Clashes with the rebels were infrequent, brief, but bitter, especially among the red-hot gullies and caves of the Ennedi and Borkou massifs which rise in dead slabs from the desert south-east of Tibesti. One officer and six men were killed, and about 20 wounded. The companies of EMT 1 returned to Corsica in April 1970; those of EMT 2 stayed until that December.

This arid, impossibly poor, but potentially mineral-rich wasteland has never known peace since independence in 1960. In 1978–79 a further round in the long-running battle between the followers of Habré and Ouedei led to légionnaires being flown in once more; but this time only in individual companies, and largely to guarantee French lives and property at N'djamena, the capital. Sub-units of the 2ᵉREP were in Tchad for a third time in 1984, during the 'phoney war' that followed Col. Gadaffi's annexing of the uranium-rich northern frontier area; and at various times since the original posting of 1969–70 the regiment has provided small training teams.

Djibouti

Scarcely more inviting in landscape and climate is the French-allied enclave of Djibouti on the 'Horn of Africa' opposite Aden. This strategic territory, nominally independent, still has a large French garrison of which the main component is the Legion's 13ᵉ Demi-Brigade. Other Legion units regularly rotate through the territory, for desert training or to increase security at times of tension, including particularly squadrons of the Legion's armoured cavalry and companies of the 2ᵉREP. In February 1976 Legion paras of the 2nd Co. took part (with the recce squadron of the 13ᵉDBLE, and a Gendarmerie unit) in the rescue of 31 children in a school bus hijacked by Somalian terrorists and driven to the Somalian border at Loyada. Apart from the kidnappers, the légionnaires faced covering fire from Somali troops on the border. Sadly, two children were killed and five wounded in the final exchange of fire; a Legion officer, and two adults on the bus, were also hurt. Seven terrorists died, plus an unknown number of Somali troops, whose fire positions were thoroughly shot up by the armoured cars of the 13ᵉ.

It was in Djibouti, and ironically in a peacetime accident, that the Legion suffered its worst losses in a single incident since the height of the Algerian War. On 3 February 1982, during a parachute exercise, a transport aircraft crashed killing two officers and 27 NCOs and men, mostly from the 2nd Platoon, 4th Co. of the 2ᵉREP.

Kolwezi

On 13 May 1978 a force of between 1,500 and 4,000 Katangan rebels of the 'Congolese National Liberation Front' (FNLC), crossing from Angola into Zaire's Shaba Province, captured the mining town of Kolwezi—and the 2,300 white technicians and their families who lived and worked there. Settling down to a leisurely sack of the town, the FNLC at first kept the worst of their savagery for the unfortunate Zairean inhabitants. But as the days passed, they began to maltreat the terrified white hostages—men, women and children. Swaggering hostility turned into random violence, rape, torture, and murder. Some poor wretches were dragged before lunatic 'courts', and shot for 'collaborating' with President Mobutu's government. Driven into compounds or cellars under guard, or hiding in their homes, the European and Asian civilians waited in terror as the tempo of massacre increased.

The Zairean Army was, predictably, incapable of mounting a rescue. After long hesitation Belgium, the former colonial power, declined to play more than a strictly humanitarian rôle. Through French

19 May 1978: paras of the 2ᵉREP rig up for the Kolwezi jump on the airfield at Kinshasa, Zaire. To save time and weight they had flown from Corsica the previous day without 'chutes, and American T-10s were provided at Kinshasa. The légionnaires found that much of their jump equipment—heavy weapons valises, cargo bags, etc—was incompatible with the harness fittings of the T-10, and they had to rig their rucksacks and weapons as best they could with cord and tape. However, the big T-10 canopy did give them a softer landing amid the bush and concrete-hard termite hills of DZ Alpha. Note company-colour triangle and platoon number on rucksack.

Lt. Col. Erulin photographed during the later stages of Operation 'Léopard', after changing from 'Satin 300s' into camouflage fatigues. Note that these are the *'toutes armes'* **pattern, and lack 'airborne' features at neck, pocket and cuff: cf. Plates F2 and G3, and the photos accompanying the Plates commentaries. This photo shows the webbing harness clearly; the quick-release belt buckle appeared in about 1957. The large 'commando' first aid pouch is taped to the left suspender. Note Erulin's rank patch, with alternating gold and silver** *galons.*

mortar and recce platoons flew 6,000 km to Kinshasa in French DC-8s. (The second echelon, with the regiment's vehicles, would later fly direct to the Shaba provincial capital, Lubumbashi, on US C-5 and C-141 transports.) By 11.30 pm on the stifling night of the 18th, after two days and a night without sleep, Erulin and the first of his men were on the ground at Kinshasa, struggling to improvise a battalion combat jump at short notice and with inadequate facilities.

Four C-130s and one C-160 were available for the drop. With their weapons and equipment hastily rigged to the harness of unfamiliar American T-10 'chutes, and crammed 80 men to a C-130 meant for 66, 405 légionnaires took off at about 11.30 am on the 19th. Red-eyed with exhaustion, and carrying only the minimum of kit, they faced an enemy of unknown strength on an unreconnoitred DZ; and knew that they would have to cope with whatever they found for at least three days before expecting support of any kind. After another four hours in the air the reduced HQ and 1st, 2nd and 3rd Cos. tumbled out the doors over Kolwezi at mid-afternoon on 19 May – almost exactly 24 years after the unit's last battalion combat drop over Dien Bien Phu. . . .

The drop on to 'DZ Alpha'—an expanse of elephant grass and tall termite hills at the north-east corner of the Old Town—was scattered, but met little ground fire. While the 3rd Co. held the bridge leading from the New Town to the east, the 1st headed south towards the Jean XXIII school some 1,000 m. away; and the 2nd pushed west towards the hospital and Gécamines complex, about 2,500 m. from 'Alpha'. Both companies had a confused, scrambling fight against dispersed groups of FNLC as they pushed through a maze of alleys, shanties and patches of scrub. Their determination to release the hostages before they could be massacred was fuelled by the dreadful sights which they came across every few moments: fly-blown corpses, black and white alike, lying everywhere in the streets and gardens. The paras took risks unthinkable against a more organised enemy, and made rapid progress. By nightfall the main objectives had been reached; an attack from the New Town by three armoured cars had been beaten off with loss; and the first hostages, hysterical with relief, were being escorted to the Jean XXIII school. Huddled in ponchos

representatives in the capital, Kinshasa, President Mobutu appealed to President Giscard. And early on the morning of 17 May a signal reached Calvi, warning Lt. Col. Erulin that his regiment was on six hours' notice of movement.

To prevent a wholesale massacre of hostages, speed was vital; and it is the speed and improvisation of Operation 'Léopard' which remain most impressive. By 8 pm on the 17th Erulin had somehow assembled the personnel of his regiment, who had been scattered on normal duties. At 1.30 am on the morning of the 18th the movement order arrived. By 8 am the bulk of the 2ᵉ REP were at Corsica's Solenzara airport. During the 18th, 650 men of the tactical HQ, four rifle companies, and wherever night found them, facing their third night without sleep on a diet of dexadrine, the paras beat off scattered attacks by wandering groups of FNLC 'Tigers' throughout the hours of darkness.

Dawn brought the second wave of transports, and 4th Co., the mortars and the recce platoon joined the fight. During 20 May the 1st and 2nd Cos. finished clearing the south and west quarters of the Old Town; 2nd then went to support 4th, who got into a fierce firefight in the northern Metal-Shaba suburb after rapidly clearing the New Town—with help from the mortars and recce troopers, some 80 'Tigers' were killed. The 3rd Co. pushed south to the Manika estate; and Erulin, from his HQ in the horribly bloodstained Impala Hotel, was able to organise the first hostage evacuation via the nearby airstrip, where Belgian troops and medical teams had now landed. Amid only sporadic firing, some paras were at last able to snatch some sleep on the night of 20/21 May, though the sights they had seen did not make for sweet dreams: in one charnel-house they had found 38 men, women and children lying stiff in their own dried blood.

On 21 May the vehicles arrived from Lubumbashi. Until 28 May the regiment carried out a wide-ranging programme of patrols over a 300-kilometre radius. There were sharp clashes at Kapata and Luilu. On the 28th the 2ᵉREP pulled out for Lubumbashi; by 4 June they were back in Corsica. They had lost five dead and suffered 25 wounded; killed about 250 FNLC, and captured 163; accounted for two armoured cars, four RCLs, 15 mortars, 21 rocket launchers, ten MGs and 275 small arms; and saved more than 2,000 innocent civilians from cruel deaths.

The Kolwezi operation showed just what 'quick reaction' units can achieve. A hundred rules of prudent soldiering were ignored, and the gamble paid off in lives saved. The world now saw what the 2ᵉREP could achieve under even the most unfavourable circumstances; and for the first time the French people as a whole realised that their government had a weapon which could strike with ruthless violence over enormous distances and at great speed. The Kolwezi jump brought the 2ᵉREP its seventh citation in Army orders.

Beirut

In complete contrast, the regiment's CCS, 1st and 3rd Cos. carried out a highly sensitive mission in August–September 1982 with notable restraint. They were among elements of the French 9th Marine and 11th Parachute Divisions sent into the murderous cauldron of besieged Beirut as part of the first Multi-National Force, to supervise the agreed withdrawal of Arafat's defeated PLO. Alongside men of the 3ᵉRPIMa (formerly 3ᵉRPC), 17ᵉRGP and RICM, they took over the port area and patrolled surrounding quarters of the ruined

A section commander ('chef de groupe') or team commander ('chef d'équipe') of the 2ᵉREP in the port area of Beirut, early September 1982; junior leaders are distinguishable by binoculars and TR-PP-11B 'handie-talkie' radios. He wears the Mle 1974 webbing issued to the regiment in 1980, and a national brassard; and carries the FAMAS assault rifle.

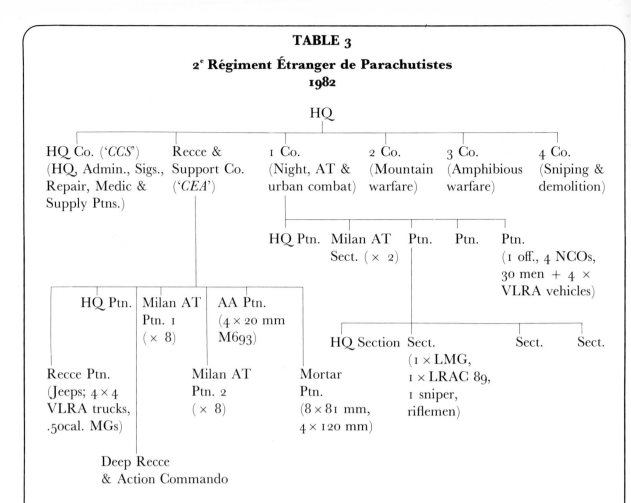

TABLE 3
2ᵉ Régiment Étranger de Parachutistes
1982

HQ

- HQ Co. ('CCS') (HQ, Admin., Sigs., Repair, Medic & Supply Ptns.)
- Recce & Support Co. ('CEA')
- 1 Co. (Night, AT & urban combat)
- 2 Co. (Mountain warfare)
- 3 Co. (Amphibious warfare)
- 4 Co. (Sniping & demolition)

1 Co.:
- HQ Ptn.
- Milan AT Sect. (× 2)
- Ptn.
- Ptn.
- Ptn. (1 off., 4 NCOs, 30 men + 4 × VLRA vehicles)

Ptn.:
- HQ Section
- Sect. (1 × LMG, 1 × LRAC 89, 1 sniper, riflemen)
- Sect.
- Sect.
- Sect.

Recce & Support Co.:
- HQ Ptn.
- Milan AT Ptn. 1 (× 8)
- AA Ptn. (4 × 20 mm M693)
- Recce Ptn. (Jeeps; 4 × 4 VLRA trucks, .50cal. MGs)
- Milan AT Ptn. 2 (× 8)
- Mortar Ptn. (8 × 81 mm, 4 × 120 mm)
- Deep Recce & Action Commando

Notes:

Total strength, c.1,275 men; the CEA alone has 237 men and 77 vehicles. In motorised rôle, all combat and support elements have integral transport: French-built jeeps, and VLRA 4 × 4 patrol trucks mounting .50cal. MGs. All companies, incl. CEA, trained and exercised as parachute infantry. Additional specialist company tasks as indicated: the company maintains a strong cadre of qualified specialists, and there is much cross-training between companies. The 'Deep Recce Cdo.'—*Commandos de Renseignements et d'Action dans la Profondeur*—are highly trained 'pathfinder' parachutists skilled in intelligence gathering and other missions behind enemy lines. (This sub-unit was formerly termed 'SOGH'—*Sauteurs Opérationnels à Grande Hauteur*.)

Lebanese capital. Though never less than firm, and ostentatiously ready for any trouble, the paras completed their mission without serious incident.

* * *

The 2ᵉREP, whose impressive establishment and firepower is indicated in the accompanying Table 3, remains today one of the West's most powerful and flexible rapid intervention units. In these days of mass unemployment, the Legion as a whole is able to reject seven out of ten would-be recruits. Competition to pass the demanding selection training for the para regiment is even stiffer. Trained to a razor's edge, the 2ᵉREP earns its extra pay—about a third higher than in other Legion

1: LMG loader, 1[er] BEP; Tonkin, 1950
2: LMG gunner, 2[e] BEP; Ba Vi, Jan. 1952
3: Sgt., 1[er] BEP; Phu Doan, Nov. 1952

A

1: Sgt., 2ᵉ BEP; Ba Vi, Dec. 1951
2: NCO, 2ᵉ BEP; Red River Delta, Apr. 1952
3: Capt. Cabiro, 1ᵉʳ BEP; Dien Bien Phu, Nov. 1953

B

1: Soldat, 1ere CIPLE; Hanoi, Apr. 1952
2: Capt., 1er BEP; Hanoi, May 1951
3: Soldat 1ere cl., 2^e BEP; Hanoi, Nov. 1951

C

1: Adjudant, 1^{er}REP; Algiers, Oct. 1957
2: Lt. Col. Jeanpierre, 1^{er}REP; Algeria, 1957-58
3: LMG ammo carrier, 2^eREP; Algeria, 1959-60

D

1: Lt.,2^{e}REP; Algeria, 1958
2: Capt., 1^{er}REP; Paris, 1961
3-8: Rank and branch insignia –
   see Plates commentaries

E

1: LMG gunner, 2^{e}REP; Tchad, 1969-70
2: Sgt. section leader, 2^{e}REP; Djibouti, 1979
3: Cpl.-chef team leader, 2^{e}REP; Tchad, 1984

F

1: Capt., 1^{er}Cie, 2^{e}REP; Kolwezi, May 1978
2: Gren.-volt., 2^{e}Cie, 2^{e}REP; Kolwezi, May 1978
3: Tireur d'élite, 4^{e}Cie, 2^{e}REP; Kolwezi, May 1978

G

1: Gren.-volt., CCS, 2ᵉREP; Beirut, Sept. 1982
2: Gren.-volt.,3ᵉCⁱᵉ 2ᵉREP; Beirut, Sept. 1982
3: Cdt., CCS, 2ᵉREP; Beirut, Aug. 1982

H

1: Cpl., 2ᵉREP, 'tenue de tradition'; Calvi, 1982
2: MP sgt., 2ᵉREP; Calvi 1978
3: Gren.-volt., 2ᵉREP; Corsica, 1984

I

Unit insignia – see Plates commentaries

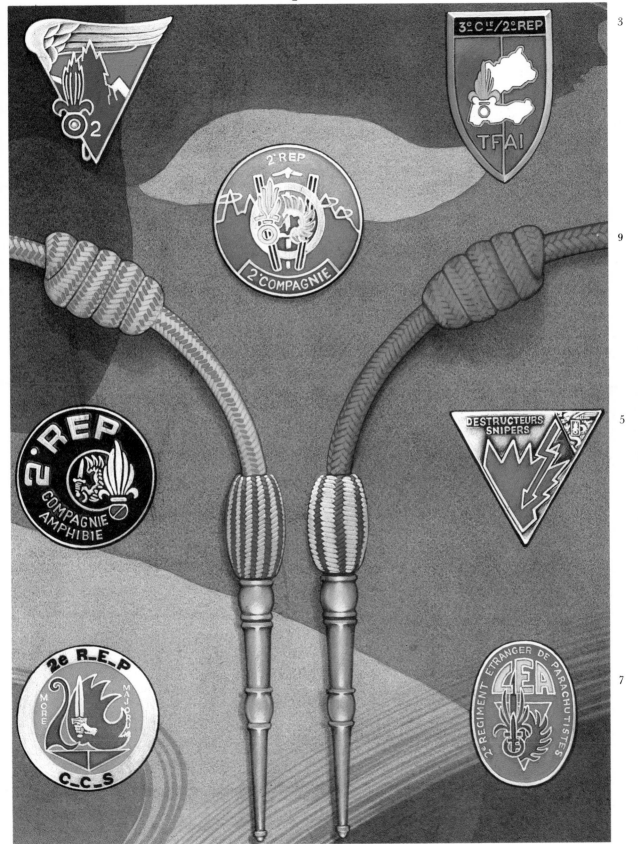

Unit insignia – see Plates commentaries

K

1

3

Formation insignia – see Plates commentaries

regiments—in full measure. Since France shows no sign of renouncing her commitments to her former colonies in sub-Saharan Africa and elsewhere; and since the Third World shows no sign of renouncing rebellion and civil war as everyday instruments of political life; so the 2ᵉ REP still offers the légionnaire his best chance of seeing active service during his five-year enlistment.

The Plates

Authors' note: When describing, e.g., sequences of insignia, we have followed French convention in referring to ranks below *caporal-chef* as 'troops'. We describe as *sous-officiers* or 'NCOs' the ranks of *sergent*, *sergent-chef* and *sergent-chef-major*; and although there have been some inconsistencies, in point of practice the *caporal-chef* may be regarded as an NCO, at least since the beginning of the 1960s. We use the British term 'warrant officers' to describe *adjudant*, *adjudant-chef* and *major*.

Note that the term 'khaki' is used here in its European meaning: a drab brown, similar to the actual shade of US Army wool service dress of the 1940s, and as used for generations by the British and French Armies for 'battledress' and service uniform. The colour Americans call 'khaki' or 'suntan' is termed here 'khaki drill'.

A: Indochina, 1950–54
The French Far East Expeditionary Corps presented a makeshift appearance, particularly in the years before the increase in US aid in 1950. Defeated in 1940, occupied and plundered, then violently liberated in 1944, France was an almost ruined nation; her army had to make do with what it could find. The bulk of the French forces had been re-equipped by the United States in 1943–45; and significant amounts of British materiel were also in use. The gradual addition of new French equipment to this motley array resulted in the Foreign Parachute Battalions wearing at least five different main types of combat clothing during the Indochina campaign, in a bewildering number of combinations. The mixture of mainly American webbing equipment with some British items, and later with the first issues of slightly differing French

US Army 1944 camouflage-printed herringbone twill field jacket, as used by the BEPs in Indochina (among several other models). See background to Plate J for outside scheme. Note the 'gas flap' at the neck; and the paler version of the scheme, in two tones of light brown on a pinkish beige background, visible on the inside surfaces. The jacket is not reversible, however: the pockets are not accessible from the inside, and the collar is not lined with the inside scheme. (Simon Dunstan)

equivalents, add further possibilities for confusion.

In the late 1940s all French troops used a fairly random mixture of French, US and British khaki drill clothing, and British and Indian Army jungle green drill. By 1950 camouflage uniform was the norm in the parachute units; this was initially drawn from two sources.

The US M1942, M1943 and M1944 two-piece camouflage-printed herringbone twill fatigues were provided in quantity. There were at least one US Army and three US Marine models in use, displaying a considerable variety of detail differences between basically similar garments made from the US camouflage material: these usually showed in the pockets, front closures and collars. The whole uniform was sometimes seen in both BEPs, perhaps more frequently in the 2ᵉ than the 1ᵉʳ: but more often the jacket alone was worn, in combination with trousers of another pattern.

Large stocks of the British 1942 camouflaged windproof smock and over-trousers were also acquired. (This is the set widely, but incorrectly, termed 'SAS windproof'.) Issued to parachute units

Trousers of the American camouflage set, with one leg rolled back to show inside scheme. Just visible on the other leg are buttons and a tightening tab above the ankle. Note the large gusset inside the flies. (Simon Dunstan)

detail: pocket closures, tightening tabs at wrist and hip, etc. Photos show different versions promiscuously mixed together in the ranks. At least two models were worn by BEP personnel, usually reserved for parades: the 'tenue de saut d'Extrême Orient', which had darker shoulder reinforcement and internal skirt pockets; and the 'tenue de saut Mle.47/51', retrospectively designated 'modèle coloniale'. A streaky camouflage version of the '47/51' was issued from spring 1953, and a modified '47/52' camouflage suit that autumn. (For the only published classification of French para uniforms known to us, see D. Lassus, Militaria magazine, nos. 6, 7 and 8.)

The usual field headgear throughout the war was the French stitched-fabric bush hat in khaki drill, green drab, or camouflaged material. This had a continuous cloth chin tape; and a snap fastener to hold the right brim up against the crown, Aussie-fashion. Often the tape was knotted over the crown to hold both sides of the brim up, 'cowboy'-style.

The issue steel helmet was the US M1, sometimes locally modified by the addition of a rear attachment bracket so that crossed 'paratrooper' chin straps could be fitted. Large numbers of the standard, unmodified M1 were also used; and smaller numbers of the M1C paratrooper's version, with extra webbing A-straps and leather chin cup.

The Legion paratroopers are reported to have agonised over the choice of a beret colour. (For practical reasons the traditional képi had to be replaced by a beret for front-line use—although it is reported that some officers had their cherished képis carried on drops by their orderlies, and donned them on the DZ!) As the corps colours are green and red, and as both colours had been chosen for berets by other corps (respectively, the French Navy's Commandos and the Colonial Paratroopers), some wits suggested that the Legion paras should wear a halved beret, or one in zigzags or concentric rings of the two colours. . . . Green was finally selected, to match the primary insignia colour of the Legion.[1] Worn from early in 1949, the beret is not much seen in photos before about 1951. Photos show that in a number of individual cases the calot was preferred by junior leaders in both BEPs: this was the green

under the title 'Survêtement '42' (or more popularly, 'sausage skins'), it was sometimes but not often seen in the BEPs as a complete suit. More usually the trousers were worn in combination with the US camouflage jacket. The hooded, pull-over smock of the British set was often retailored locally, frequently losing the hood in favour of a shirt-type collar, and/or being opened down the front and fitted with a zip, buttons or snap fasteners. The trousers sometimes acquired extra pockets.

The late 1940s saw the arrival in quantity of the new standard French Army combat fatigues, 'treillis de combat Mle 1947'. While it lacked any specifically 'airborne' features, this two-piece olive drab suit was often worn by the BEPs, either complete or in combination with other types.

Early in 1950 the first of a bewilderingly complex series of French-designed 'jump uniforms' appeared. Made in 'kaki foncé' (a drab tan), they progressively acquired many small differences of

[1]The beret was not officially authorised until July 1957, previously having only semi-official status as a tolerated 'coiffure de repos'.

British 1942 windproof smock, used to some extent by the BEPs in Indochina. See Plates A1 and A2, and background to Plate K. (Wayne Braby)

sidehat with a red top gusset, worn by the Legion's RMLE in North-West Europe, 1944–45.

The boots were a mixture of US Army M1943 'double-buckle' combat boots, sometimes with an extra rubber sole fitted; and conventional para 'jump boots'. In both cases the original stocks were American, but French-made or locally-made copies soon appeared. Like all other units of the Expeditionary Corps, the paras often made use of the practical 'patauga' patrol boots of canvas and rubber.

To detail every known variation of webbing equipment would be as tedious as it is pointless. The norm was the US M1912 pistol belt and associated suspenders, though some M1923 rifle belts were also issued. The double magazine pouches for the M1 carbine were very widely used, and the US water canteen and carrier were almost universal. The Thompson SMG magazine pouch was quite widely seen even after the general replacement of that weapon by the MAT 49, as French-made SMG pouches were not immediately available. British 1937-pattern web belts, 'small packs' and 'universal pouches' are seen in some photos, though not usually after about 1952. Pistol holsters were either

of webbing, closely resembling the snapped-flap British type; of leather, in French designs; or imported—there was some use of the Colt 1911 automatic and its holster. The beginnings of standardisation came with the new French 'equipement de toile Mle 1950 TAP' ('for airborne troops'), which, apart from buckle details, closely resembled the American belt and suspenders. But the different items of this set were introduced gradually, and a motley selection of US and French webbing was still to be seen in the early days of the Algerian War.

The small arms used initially were the CR 39, a folding-stock version of the old bolt-action 7.5 mm MAS 36 rifle; the .45 in. Thompson and, to a small extent, the 7.65 mm MAS 38 sub-machine guns; and the 7.5 mm FM 24/29 light machine gun. By 1950 quantities of the folding-stock M1A1 carbine were becoming available, but this never completely replaced the CR 39. The MAT 49 sub-machine gun replaced other types steadily from its introduction in the same year. Support elements used the Browning .30cal. machine gun; the 2 in., 60 mm and 81 mm mortars; and the M18 RCL (recoilless rifle) in 57 mm.

A1: Chargeur FM, 1ᵉʳ Bataillon Étranger de Parachutistes; Tonkin, 1950
The 'no. 2' of a squad light machine gun team,

taking a box magazine for the FM 24/29 from the 'special haversack Mle 1924' in which he carries eight of them. He wears the US M1 helmet; the British windproof camouflage smock; the trousers of the French 1947 olive drab fatigue set; the US webbing rifle belt and suspenders; and US 'double-buckle' combat boots with added rubber soles. He carries the CR 39 rifle and OF 37 blast grenades.

A2: Tireur FM, 2ᵉBEP; Ba Vi, January 1952
Squad LMG gunner in the classic Legion para combat dress of the middle years of the war: US camouflage jacket; British camouflage trousers; US webbing, helmet and boots. Slung behind the body is the 'special musette Mle 1924' holding six magazines for the FM 24/29, or three plus the FM tool kit. This is the soldier of the RC 4 and Hoa Binh campaigns.

Trousers of the British 1942 windproof camouflage suit, used by the BEPs in Indochina. Note that there is no opening fly; the generously cut waist is tightened by a draw-tape. (Wayne Braby)

A3: Sergent, 1ᵉʳBEP; Phu Doan, November 1952
Several photos taken during Operation 'Marion' show 1ᵉʳBEP paras wearing complete 1947 French fatigue suits with bush hats. The crotch-length jacket has a shirt-type collar with long upper points; four internal pockets with external, single-point flaps; and all buttons concealed. As well as conventional side and hip pockets, the trousers have large 'bellows' cargo pockets on the outer thighs, their straight flaps fastened by two concealed buttons. Buttoned 'storm tabs' were fitted at wrists and ankles. This NCO wears US webbing, and carries the M1A1 carbine. Most junior leaders also carried sidearms, in chaotic variety until the French MAC 50 9 mm automatic became standard. The rank chevrons and branch of service *écusson* normally worn on the left sleeve of parade and walking-out dress (see under Plates E3-E8 for detail of all insignia) are here pinned to the left breast pocket, reversed for a better view of the ranking.

B1: Sergent, 2ᵉBEP; Ba Vi, December 1951
A composite, from two photos. One shows Sgt. Frouart of the 2nd Co. wearing this US jacket with the original USMC stencil above the temporarily attached ranking and *écusson*. (*Errata:* The breast pocket should have no flap.) From another photo we take the US camouflage trousers, and the belt and SMG magazine pouch of the French 1946 leather equipment set—leather infantry equipment was occasionally, though not often, worn by paratroopers. He holds a MAT 49; and stands by the M18 57 mm RCL used by company HQ platoons and battalion HQ Co. 'heavy platoons'. The radio is the American SCR 536.

B2: Sous-officier, 2ᵉBEP; Luc Dien, Red River Delta, April 1952
This '*tenue de saut Mle.1947*' was in khaki drab—it has printed far too green here. It had a deep 'collar'; tabs buttoning backwards at the hips, forwards on the cuffs; two-snap breast pockets; two-snap, inward-slanting skirt pockets; slanted, buttoned, internal rear skirt pockets; and a Denison-type 'beaver tail'. The trousers had two-snap, pleated cargo pockets. Individual examples were photographed, but most BEP personnel retained US and British camouflage uniforms. M1A1 carbines, initially issued to junior leaders, were more

Front and rear of the French 47/53 camouflaged smock for airborne troops, in this case bearing para wings and a 1^{er}REP badge. Note collarless neck; left chest pocket with small patch pocket on face, and zip in inside edge; slanted skirt pockets; and crotch-piece snap-fastened up behind when not in use. (Wayne Braby)

widespread in the ranks by *c.*1953. Note webbing medical pouch of French 1950 equipment worn on belt.

B3: Capitaine Cabiro, 1^{er}BEP; Dien Bien Phu, November 1953

The next suit to appear was the '47/51', also in dark khaki, later made in streaked French camouflage of tan, light red-brown and green. It differed from the '47' in having three-snap breast pockets; forward-buttoning hip tabs; two zipped, vertical, 24 cm ventilation slits at the sides of the smock back; and three small, buttoned, pointed-flap bellows pockets on the thigh fronts—two 'inboard' of the cargo pockets and one high on the left hip front.

Photos suggest that by November 1953 the BEPs and CEPML had the '47/52' camouflaged suit, which reached units from autumn 1953. Colour photos taken at Dien Bien Phu by Lt. Rondy of the 1^{er}BEP show it worn with similarly-camouflaged bush hats. This figure is based on a photo of the legendary 'Cab', CO of 4th Co. during Operation 'Castor'.

Note features common to the '47/51' and '/52': extra patch pocket on face of left breast pocket, and vertical zip down its 'inboard' edge; six press studs for fastening the 'tail' between the legs, between the slanted skirt pockets; and 'collarless' neck, opened into two triangular 'lapels', cut smaller than on the 'Mle.47'. The '47/52' lacked the rear skirt pockets and ventilation slits, and had (hidden here) buckled cuff tabs, though retaining buttoned hip tabs. The trousers had the three small frontal pockets, though the upper left one is hidden here by the smock skirt.

Later modifications included the '47/53' (tabs on cuffs buttoned; on hips, buckled); '47/54' (no 'tail'); and '47/56' (no tightening tabs; elasticated, buttoned cuffs; drawstring hem). The three small trouser front pockets disappeared from the '/53' suit onwards.

French 'Mle.1950 cloth equipment for airborne troops' is distinguishable by suspender buckle details; some versions also used British '37-type belt buckles. By 1953 officers usually carried carbines. Fighting knives or carbine bayonets slung on the left thigh were very common throughout the life of the BEPs and REPs. Cabiro has a webbing mapcase slung on his belt, and holds the handset of the US

AN/PRC 10 radio.

The pre-1953 version of the dark green beret was made of two pieces, and had a visible crown seam. It would be worn here with the silver badge common to Metropolitan and Foreign parachute units: a circle enclosing a winged fist brandishing a short sword. (Elsewhere, the author has incorrectly stated that in recent years the 2ᵉREP wore a gold version: although it contradicts normal French branch of service 'lace and metal' conventions, it is now clear that in fact the Legion paras have always worn a silver cap badge. And that the *ancien légionnaire* who swore to the contrary was having a momentary lapse. . . .) The only other insignia worn are the three gold *galons* of a captain, on a midnight blue chest patch; shoulder strap loops were also common on combat dress.

Képi **of a Foreign Legion** *sous-lieutenant*—**see detailed description in commentary to Plate C2.**

C1: Soldat 2ᵉclasse, 1ᵉʳᵉCompagnie Indochinoise Parachutiste de la Légion Étrangère, 1ᵉʳBEP; Hanoi, April 1952

In April 1951, as part of Gen. de Lattre's programme for training local troops by integration within French units, each Legion battalion was ordered to form a Vietnamese company, and each company a Vietnamese platoon. The 1ᵉʳᵉ and 2ᵉCIPLEs, attached to the 1ᵉʳ and 2ᵉBEPs respectively, soon proved themselves in action. (According to Bernard Fall, by the time of Dien Bien Phu the 1ᵉʳBEP had no less than 336 Vietnamese soldiers out of a total strength of 653 all ranks.)

The summer parade and walking-out dress of the Legion was a shirtsleeve uniform of khaki drill, the material having a slight pink/brown cast when compared with British and US equivalents. It was embellished for parades with the Legion's *epaulettes de tradition* and *ceinture bleue*; occasionally units or parts of units, e.g. colour parties and honour guards, wore whitened webbing or French 1946 leather equipment—in this case, the US pistol belt and leggings.

The white-covered *képi*, a cherished item of Legion 'folklore', was not issued to Vietnamese personnel: they had not fulfilled all the traditional enlistment conditions, and would in any case have required awkwardly large numbers of very small sizes. Instead they wore (for parade and walking-out only) a white beret, probably white-dyed

examples of the French general service khaki drill beret, with the usual silver paratrooper's badge. The rear tightening tapes, left hanging in the French manner, were—in some cases at least—coloured Legion red and green. (This conformed with practice in other units of the fledgling Vietnamese Army.) At other times the Indochinese wore the green beret.

On the left sleeve, obscured here, the Legion's branch of service *écusson* was worn, temporarily hooked on, as were all cloth insignia worn on shirtsleeves; see Plate E3–E8 for details of such insignia, and of the sleeve ranking also worn on this uniform. The parachutist's qualification wings ('bicycle badge') are pinned to the right breast; and the 1ᵉʳBEP's metal unit insignia is attached below them, pinned to a leather fob in the usual French manner. The citation lanyard or *fourragère* is carried across to the top shirt button for parades. The 1ᵉʳBEP won its second citation on 25 January 1951, thus qualifying for the lanyard of the *Croix de Guerre*; in Indochina this meant the *C de G 'des Théâtres d'Opérations Extérieurs'*, in pale blue flecked with red.

C2: Capitaine, 1ᵉʳBEP; Hanoi/Bach Mai, May 1951
Normally the retiring commander of a Legion unit hands the flag to his relief during a formal ceremony. The annihilation of the 1ᵉʳBEP in the RC 4 battles, and the commitment to combat of the re-formed battalion just five days after it disembarked in Indochina in March 1951, prevented such a ceremony. It was on 11 May 1951 that the unit paraded at the Bach Mai rear headquarters; and Gen. de Lattre, the C-in-C French Indochina, presented a *fanion* to the new CO, Capt. Darmuzai, and simultaneously decorated it with the *C de G TOE* lanyard marking the citation of 25 January.

The photo on which we base this figure shows a *fanion* divided into the Legion's green and red, mounted in the usual way on a 'ramrod' in the muzzle of a rifle, with a brass finial in the form of the Legion's seven-flamed grenade. It bears an embroidered silver and gold paratrooper's cap badge motif beneath a large '1', with the words 'BATAILLON/ÉTRANGER' in two arcs and 'DE PARACHUTISTES' in a line. (The 1ᵉʳBEP *fanion* from Indochina preserved today at Aubagne is rather different, and may date from slightly later.)

The officer wears the khaki drill shirt of summer uniform, with the trousers of the French 1946 battledress uniform worn for winter parades and walking-out. The shirt bears the officer's shoulder boards, and would also display the Legion *écusson* on the left sleeve (again, see Plate E3–E8 for details). Below the parachutist's *brevet* the unit badge—as frequently seen in Indochina—is pinned directly to the shirt. The medals are notional, but typical for an officer who had seen combat with the French Resistance and Free French Forces.

The outfit is completed by a whitened US pistol belt, a leather holster of French design, 'jump boots', and the officer's *képi*. All commissioned and

French 47/56 camouflaged smock for airborne troops; the pocket flaps are shown raised here, to display the three pairs of snap fasteners. Note elasticated, buttoned cuffs; drawstring in hem; pale green lining—the same colour as the background to the camouflage scheme, shown in detail on Plate L; and buttoned fly covering a half-length zip. The skirt pockets are set on the smock straight, not slanted. (Simon Dunstan)

warrant officers wear it in midnight blue (in practice, black) with a red top edge and top surface. It has a gold false chinstrap, gold buttons, and a gold-embroidered Legion grenade. Around the top of the body, lines of metallic Russia braid follow the rank/number sequence listed in the commentary to Plate E6, including the alternating silver and gold of *lieutenant-colonel*. (Warrant officers wear silver or gold lace interwoven with red V-shaped threads, both here and on the top and top edges of the *képi*.) At front, back and each side short lengths of braid pass vertically from the top rank *galon* up and over the top edge of the cap: one for warrant officers, *sous-lieutenant* and *lieutenant*; two for *capitaine*; and three for all field ranks. The top surface is surrounded by a single edging of lace, and bears a lace quatrefoil, the number of braids corresponding to the number of vertical braids at the top edge.

C3: Soldat 1ᵉʳᵉclasse, 2ᵉBEP; Hanoi, November 1951
This member of an honour guard inspected by Gen. de Lattre shortly before his departure wears the winter parade and walking-out uniform, with parade embellishments as Plate C1. The 1946

battledress, worn by légionnaires for cold weather parades from 1947 until its withdrawal in 1980, resembles the British 1949 pattern in all but minor details. It has an open notched collar; breast pockets with box pleats and flaps *en accolade*; plain shoulder straps; brass uniform buttons; and plain trousers (i.e. without the large left thigh pocket of the British original). In Indochina it was often worn, even for parades, with the shirt collar pressed open. Sleeve *écussons* and ranking were worn—here the single chevron of private first class (see commentary, Plate E3–E8). This belt is a whitened example of the French 1946 leather type. (This soldier would almost certainly wear the BD trousers tucked into US web leggings, whitened, and perhaps hidden for half their depth by the deep trouser 'pull-down'.) The MAT 49 is carried here with the magazine pivoted forward. Parachutist's and battalion insignia are worn on the right breast; on the left, personal awards of the *C de G TOE* and the Wound Medal. (The battalion did not receive its collective award of the *C de G TOE* lanyard until the 21st of that month.)

Until the early 1960s the ranks below *caporal-chef* were issued a headgear which never, in practice, saw the light of day: a *képi* in midnight blue and red, with dark blue piping as on the NCO's type illustrated in Plate I2, but with a black chin strap and a red grenade badge. This cap was worn with a white cotton cover by privates and corporals in all orders of dress. For parades a second black chin strap was fitted, and worn beneath the chin.

D1: Adjudant, 1ᵉʳRégiment Étranger de Parachutistes; Algiers, October 1957

Several slightly differing models of camouflage fatigues appeared during the mid- to late 1950s; the differences did not go much beyond details of e.g. the tightening tabs at hip and wrist. (See notes under Plate B3; and—once more—Denis Lassus's masterly classifications in *Militaria* magazine nos. 6, 7 and 8.) The 47/56 type, illustrated here, differed in having the smock skirt pockets set straight rather than slanted. By this time, too, the trousers had lost the three small frontal thigh pockets, and had acquired vertical 'letterbox' flaps over the pockets in the side seams, held by three snap fasteners. Camouflage clothing rarely carried insignia when the troops were on operations; but senior ranks

REP para in Algeria, 1959, wearing the green, collarless, quilted wind jacket over his camouflage fatigues.

displayed ranking, normally on chest patches or as metal shoulder strap clips—here we show the slightly old-fashioned shoulder strap loops in stiff fabric. Photos do show the occasional use of formation patches, however; and we include here that of Gen. Massu's 10ᵉDivision Parachutiste.

For most operations the green beret was worn; when serious fighting was anticipated, as during the 'battle of the frontiers' with battalion-size ALN units, the REPs carried the steel helmet, usually with a net cover. The US M1 was still to be seen, alongside the French 1951, or its cross-strapped 1956 paratrooper's variant. It should be stressed that, alone of parachute units in Algeria, the REPs never adopted the camouflaged 'Bigeard cap'.

Minimal belt equipment was worn for urban security duties: usually only the belt, suspenders, pouches, combat knife, and a canteen. Officers and warrant officers usually wore only the MAC 50 automatic, here carried in a second pattern of webbing holster with a deep, strapped flap; but this veteran platoon second-in-command, clearly anticipating a stiff firefight in the chaotic stone maze of the *casbah*, has taken a MAT 49 and one quadruple magazine pouch. This sturdy 9 mm blow-back weapon was the standard French SMG for more than 30 years. Troops usually carried two quadruple pouches for the 32-round box magazines.

D2: Lieutenant-Colonel Jeanpierre, 1ᵉʳREP; Guelma sector, Algeria, 1957–58

The legendary commander of the 1ᵉʳREP in Algeria worked his way up from the ranks by merit, and graduated from officer school in 1936. He served thereafter exclusively with Legion units, apart from a period with the French Resistance in 1943–44 which landed him in Mauthausen concentration camp. He sailed to Indochina with the 1ᵉʳBEP as second-in-command. He took much of the weight of command off the crippled Ch. de Bn. Segrétain during the disastrous RC 4 fighting in October 1950; and it was Jeanpierre who led the 29 survivors of the battalion out of the jungle, 'looking like Christ down from the cross'. Evacuated to Algeria, he returned to lead the re-formed battalion after its second annihilation at Dien Bien Phu. Passed over for immediate command of the enlarged 1ᵉʳREP, he remained as second-in-command until taking over as '*patron*' in March 1957. He spent the last 14 months of his life leading what many believe to have been France's finest fighting regiment. A rough, earthy soldier driven by a hunger for excellence, he was mercilessly demanding of himself and of his officers and men; but under his inspired leadership they achieved extraordinary results, and regarded him with awed admiration.

His quilted wind jacket and wool toque are reminders that it is entirely possible to freeze to death in deep snow in the Algerian mountains in winter; the jacket was widely issued to French forces, and was worn both over and under the combat fatigues. Various photos show Jeanpierre wearing the slightly differing variants of camouflage smock and trousers with 1950 web equipment, belt with British-style buckle, early snap-flapped pistol holster, M3 combat knife, and jump boots. His rank patch displays the alternating gold and silver *galons* of lieutenant-colonel.

D3: Pourvoyeur FM, 2ᵉREP; Algeria, 1959–60

New replacements often joined their squad's 'fire team' as ammunition carriers and support riflemen for the LMG crew—'*pourvoyeurs*'—while more experienced or immediately promising soldiers carried the MAT 49 as '*voltigeurs*' in the squad's 'assault team'. By the latter part of the Algerian War the LMG was the AA 52, whose ammunition was carried in disintegrating link belts; and the rifle was this MAS 49/56, a reliable self-loading weapon of 7.5 mm calibre with a detachable ten-round box magazine and an integral grenade launcher. (In the early years of the war the CR 39, M1A1 carbine, and even isolated examples of the Garand M1 rifle were still in use; as was the MAS 49, an intermediate design similar to the later 49/56 but with forestock, butt and muzzle superficially resembling those of the old MAS 36 bolt-action weapon.)

Photos show a random mixture of different variants of the camouflage fatigues; in the 2ᵉREP the earliest 48/51 pattern seems to have been issued quite late in the war. Note that in the early 1960s the fashion for wearing tiny, shrunken berets had not conquered the French! An exaggeratedly flat, forward-pulled shape, with the badge near the right ear, is seen in many photos. From about 1958 the usual boots were the double-buckle French '*rangers*', sometimes issued brown and overpolished black; but there were still plenty of high-lacing jump boots in use as well. The double rifle clip pouches, '*cartouchières 1950/53*', appeared in quantity from about 1954. The 'bergen' rucksack differed only in small details from that first issued to légionnaires for the 1940 Narvik campaign.

E1: Lieutenant, 2ᵉREP; Philippeville, Algeria, 1958

During the Algerian War the paratroopers' camouflage fatigues became, by an association of ideas, the symbol of military determination, even of a political attitude. The paras, of all categories, were France's most effective units, while the olive-clad conscript infantry were felt to have a somewhat ambivalent attitude to the conflict. The camouflaged '*tenue léopard*' became fashionable for all occasions—one officer even wore his to an opening night at the Paris Opéra! More conventionally, this officer wears fully badged and decorated combat dress for a regimental parade. (At the beginning of the 1960s camouflage fatigues

The 1956 smock 'badged' for parade wear with the insignia of the 1er REP. The regimental badge (see Plate J2) is pinned to the right chest pocket, and para wings above the pocket. The patch of the 10e DP (see Plate L1) is fixed to the right shoulder. The citation lanyard (see Plate K8) is carried across to the inner snap of the left chest pocket. (Simon Dunstan)

were sometimes seen with *epaulettes de tradition* and *ceinture bleue*, but this was never a general practice.)

Ranking is worn in the form of gilt metal shoulder strap clips, and the patch of the 25e Division Parachutiste is hooked to the right shoulder. Parachutist's *brevet* and unit badge are pinned directly to the smock. The medals indicate service in the last stages of the Indochina War and in Algeria: *Croix de Valeur Militaire*, with the silver star indicating a citation in divisional orders, won in Algeria; *Croix de Guerre TOE*, with two bronze brigade citations, won in Indochina; the Indochina campaign medal; the *Médaille Commémorative des Opérations de Securité et du Maintien de l'Ordre*, with clasp '*Algérie*'; and the Wound Medal, often worn with a small red enamel star on the ribbon.

The regiment's citation lanyard, in the colour of the *Légion d'Honneur* ribbon and with an '*olive*' in *C de G TOE* colours indicating an award in Indochina, is carried across to the inner snap of the left pocket for parade occasions. It marks the sixth citation, won by the 2e REP on 23 April 1954. (Photos of both REPs sometimes show the *fourragère* worn during the mid-1950s without the regulation '*olive*'.)

E2: Capitaine, 1er REP; Paris, 1961
An unusual combination of items, worn by an officer during the enquiry following the tragic

events of April 1961. The wearing of the para beret with No. 1 Winter Walking-Out Dress was perhaps a sign of the times, like the use of camouflage fatigues for parades: the *képi* would normally be worn with this uniform. This is the 1956 version of the officer's and warrant officer's service dress tunic and trousers, worn with laced shoes and, for No. 1 Dress, with white gloves. The trousers have a double dark brown side stripe 50 mm broad; warrant officers wore narrow dark brown piping. Legion collar *écussons* and shoulder boards identify branch of service and rank (see under Plate E3–E8); the green tie and officer's green waistcoat are also Legion distinctions. The ornate black belt, with gilt medallion-and-snake clasp, is normally seen only with the blue ceremonial uniform worn on certain state occasions, but is clearly shown in the original photo. The notional selection of medals shown here includes the *Médaille Militaire*, which indicates past service in the ranks—it is awarded only to troops and NCOs for conspicuous gallantry. The regiment's citation *fourragère*—here of the fancy type with a 'tour-de-bras'—is in *Médaille Militaire* colours of yellow and green, marking the 1er BEP's fourth citation on 17 December 1953.

E3–E8: Rank and branch of service insignia
(A check-list of the different insignia worn by different ranks on different types of uniform follows the individual captions.)
E3: Collar écusson Branch of service insignia of the Legion, worn by officers and warrant officers since c.1946, and by NCOs since 1960; the latter have less ornate gold grenades.
E4: Left sleeve écusson Branch of service insignia of the Legion, worn by troops and NCOs since c.1946, and under some circumstances by officers and warrant officers. Green grenade for troops; gold grenade for NCOs; ornate gold grenade for officers and warrant officers. For a while in the 1950s–60s, *caporal-chef* was marked by a yellow thread grenade; and in the same period regimental numbers were often shown in the grenade's lower 'bomb'. In Indochina in the late 1940s some early examples worn by officers, cut down from collar *écussons*, had only the three green chevrons rather than complete triple edge piping.
E5: Sleeve ranking; and re-enlistment chevron Here, the left sleeve presentation of a *caporal-chef*, early 1960s,

with one completed five-year engagement. The ranking system has been worn since 1946 by all troops and NCOs, the re-enlistment chevron since 1948.

Chevrons are worn on both upper sleeves as follows: *soldat 1ere classe*—one green; *caporal*—two green; *caporal-chef*—one gold above two green; *sergent*—two gold; *sergent-chef*—three gold; *sergent-chef-major* (rank discontinued in late 1960s, and rarely awarded, as a unit appointment, before then)—four gold.

Re-enlistment chevrons, worn below the *écusson* on the left (and, during the 1950s–60s, in the bend of the rank chevrons on the right) sleeve are green or gold, for troops and NCOs. (This example is in the 'intermediate' yellow then worn by a *caporal-chef*, even though the grenade is gold.)

E6: Officers' and warrant officers' shoulder boards
Branch of service and rank insignia, worn since c.1946; here, *capitaine*, but all ranks display the three green chevrons and gold grenade. Rank *galons* are as follows:

Adjudant	...	One silver stripe with red 'light'
Adjudant-chef	...	One gold stripe with red 'light'
Major (senior WO appointment since December 1975)	...	One gold stripe with red 'light', one narrow gold stripe 'inboard'
Sous-lieutenant	...	One gold stripe
Lieutenant	...	Two gold stripes
Capitaine	...	Three gold stripes
Commandant* (incl. Chef de Bataillon)	...	Four gold stripes
Lieutenant-colonel*	...	Five stripes, alternating gold/silver/gold/silver/gold
Colonel*	...	Five gold stripes

(*All these ranks have a slight separation between the 'outboard' three stripes on the shoulder board and any additional stripes.)

E7, E8: NCOs' and troops' shoulder boards Displaying branch of service and category of rank—gold grenades for *sous-officiers*, green for troops. Grad-

The medals displayed on the smock in the previous photo are the *Croix de Valeur Militaire*, and the *Médaille Commémorative des Opérations de Sécurité et du Maintien de l'Ordre*. The former, a bronze cross on a scarlet and white ribbon, bears here the silver star of a citation in divisional orders. It more or less took the place of the *Croix de Guerre TOE* during the Algerian War, being instituted in 1956; the *C de G TOE* could hardly be awarded for gallantry in a war fought not in 'exterior theatres of operations' but in what was, legally, part of France. The *Médaille Commémorative* is a gold medal on a ribbon in white and scarlet with a pale blue centre stripe; it bears here the gilt clasp '*Algérie*', and was instituted in January 1958 to mark 90 days' service on security operations. It was, in effect, the campaign medal of the Algerian War. (Simon Dunstan)

ually introduced in the early 1950s; rarely if ever seen in Indochina, but general issue in Algeria. Past service with another Legion unit may be marked by a miniature (in the 1950s–60s, sometimes by a full size) metal insignia of the old unit pinned to the end of the left hand board.

These insignia have been worn as follows:
On officers' and warrant officers' four-pocket service dress tunic:
Since c.1946, collar *écussons* and officers', warrant officers' shoulder boards.
On sous-officiers' four-pocket service dress tunic:
Since introduction of tunic c.1960, collar *écussons*, shoulder boards, sleeve ranking.
On troops' four-pocket service dress tunic:
Since introduction of tunic c.1960, sleeve *écusson*, shoulder boards, sleeve ranking.

Citation lanyard, or *fourragère*, of the type with an extra loop or '*tour-de-bras*'. **Details of the award of lanyards to the BEPs and REPs are given in the commentaries to Plates C1, C3, E1, E2, K8 and K9. There is no official significance in the use of the 'simple' *fourragère* or the '*tour-de-bras*' type—this is a matter of unit choice or availability at a particular period.**
(Below) Diagram showing the methods of wearing *fourragères*. **Thread loops extend out of the weave at the top or trefoil end, and close to the bottom end of the plaited section. A button is usually sewn to the shoulder under the outer end of the shoulder strap, though lanyards have occasionally been worn from the normal shoulder strap button.**
(A) 'Simple' lanyard. For walking-out dress both thread loops are attached to the button on the shoulder. For parade, the bottom loop is carried across the chest to the top front button of the tunic or shirt.
(B) Walking-out arrangement of lanyard with extra '*tour-de-bras*': **this is doubled, and worn on the outside of the arm, held by the shoulder strap.**
(C) Parade arrangement, with the arm passed through the doubled '*tour-de-bras*'. **(Christa Hook)**

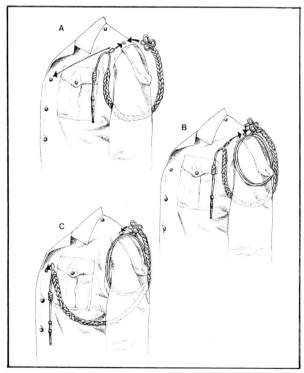

On officers' and warrant officers' battledress blouse:
Worn since late 1940s, usually only when parading with troops; sleeve *écusson*, shoulder boards. Occasional non-regulation use of collar *écussons* instead of sleeve *écusson* in early years.

On sous-officiers' and troops' battledress blouse:
Since c.1946, sleeve *écusson*, sleeve ranking; since early 1950s, shoulder boards for walking-out. For parade, c.1946 to 1980 (when BD withdrawn and replaced by four-pocket tunic), *epaulettes de tradition* replaced shoulder boards, other insignia unchanged.

(Since c.1980 the four-pocket tunic has been worn for parade by all ranks. Troops and *sous-officiers* wear *epaulettes de tradition* and sleeve ranking; *sous-officiers* wear collar *écussons*, troops wear sleeve *écusson*. Officers' uniform differs from walking-out dress only by the addition of combat boots and a pistol belt and holster.)

On shirtsleeve uniforms:
As on battledress blouse.

F1: Tireur AA 52, 2ᵉ REP; Tchad, 1969–70
The infantry squad or section ('*Groupe 801 Type Commando*') was led by a *sergent* armed with a MAT 49. It comprised an *équipe feu* or 'fire team' and an *équipe choc* or 'assault team'. The 'fire team' was led by a *caporal-chef* armed with a MAT 49, and comprised the two-man AA 52 light machine gun team, and the two-man 89 mm LRAC rocket launcher team. The 'assault team', also led by a *caporal-chef* with a MAT 49, comprised a point scout with a MAT 49, and three *grenadier-voltigeurs* with rifles and hand and rifle grenades; one might be sniper-trained, and equipped with telescopic sights.

The gas-operated 7.5 mm AA 52, fed by 250-round disintegrating link belts, replaced the FM 24/29 as the standard squad LMG during the Algerian War. The LMG version (there is also a heavy-barrel sustained fire version) has a rate of fire of about 800 rpm and an effective range of 2,000 yards.

In about 1966–67 the 2ᵉ REP began to be photographed in the olive green '*Mle 1960 tenue TAP*' shown here; apart from the colour it is virtually identical to the 1956 camouflage fatigues. This set was worn in Tchad, as also was the camouflaged 'general service' set illustrated in

The general service pattern camouflaged fatigue jacket; an early version of this was first issued to the Legion's infantry and cavalry units in 1959/60, and it is now worn by the 2ᵉREP for some overseas postings. The main differences from the old 'TAP' model are the shirt-type collar; internal chest and bellows skirt pockets all with pointed flaps and concealed buttons; an entirely buttoned fly front; and buttoning storm-tabs on the wrists. In Legion use the jacket may be seen shortened to the hips and elasticated. This example bears para wings; 2ᵉREP badge; and the 'velcro'd' rank patch of a *capitaine*—three gold *galons* on olive green. (Wayne Braby)

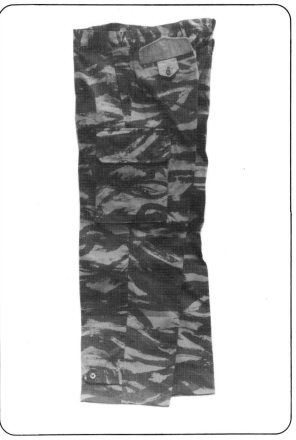

The trousers of the general service camouflaged fatigue set, showing type of fastening used on all pockets; cargo pockets without gussets; and buttoning tab at ankles. (Wayne Braby)

Plates F2 and G3. Bush hats were worn in the murderous desert climate, and a fairly exotic collection of scarves, although the traditional *cheich* is always popular. The first aid pouch, previously seen on the back of the belt, was from now on worn taped to the left suspender; and riflemen and sub-machine gunners wore on the right thigh the triple grenade pouch illustrated in Plates G and H. The LMG gunner would carry a MAC 50 pistol in a webbing holster, and a double magazine pouch. Unusually, some photos show the Legion's *écusson* and ranking worn on the left sleeve of this combat dress.

F2: Chef de groupe, 2ᵉREP; Djibouti, 1979

During frequent rotation through this uninviting corner of East Africa the paras have worn this neat and practical combination of a camouflage fatigue jacket and olive drab—or khaki drill—shorts. The

jacket is the 'general issue' type ('*treillis de combat camouflé Mle toutes armes 1947*') worn by Legion infantry and cavalry in the early to mid-1960s, though these days rather more smartly tailored than was the style in Algeria. It differs from the airborne 'TAP' version in having a shirt-type collar; a buttoned fly front; four pockets with single-point flaps and concealed buttons; and buttoning storm-tabs on the wrists. The '*rangers*' have been made in black since about 1969. Conventional webbing is worn, with two MAT 49 pouches, and the large airborne troops' rucksack. Ranking is worn on an olive drab patch fixed to the chest with 'velcro'—here, the *sergent's* two gold diagonals.

F3: Chef d'équipe choc, 2ᵉREP; Tchad, 1984

With the return of the Legion to Tchad during the crisis provoked by Col. Gadaffi's recent invasion, the paras had a chance to test under punishing

Four-pocket khaki 'Tergal' tunic of a 2ᵉREP légionnaire, 1980s, as worn for walking-out dress in winter. Note shoulder boards (see Plate E8); sleeve *écusson* (see Plate E4); 11ᵉ DP shoulder patch (see Plate L4); regimental or company badge, on a red-stitched green leather fob (see Plate J3); and lanyard (see Plate K9). (Simon Dunstan)

The same tunic, as worn for parades since 1979/80. Shoulder boards are replaced by red and green *epaulettes de tradition*, and the broad royal blue *ceinture bleue* is worn at the waist beneath the webbing or leather belt. The lanyard is carried across to the top tunic button; and medals, if any, are worn. See Plate I1. (Simon Dunstan)

conditions their new personal equipment and weapon. Since 1980 the MAS 49/56 and MAT 49 have both been replaced by the 5.56 mm FAMAS, an assault rifle of 'bullpup' design; it takes 25-round box magazines, and has a theoretical rate of fire of 950 rpm. It was also in 1980 that the new Mle 1974 webbing equipment was issued. The belt is little changed; but the suspenders are now an H-rig, with a D-ring and adjustment buckle at the front of each shoulder, and 'velcro' strips at the bottom of the rear straps, which pass round the belt and then fasten on themselves. The two pouches each take three FAMAS magazines in a 'piggy back' arrangement. (See also Plate H.)

Originally for mechanics and labourers, the *'chemise GAO'* is now popular for patrols. Well-ventilated, it still protects the shoulders from the chafing of the webbing. It fastens with three straps on each side; and some examples seem to have two open patch pockets low at the back. The usual rank patch fixes to a 'velcro' tab below the V-neck: here, the one gold and two green diagonals of *'capo-chef'*. (An alternative seen recently in Tchad is a

conventional olive drab military shirt with short sleeves. It bears a name tab on the right chest, and formation patches, in addition to the chest ranking.) The boots are apparently a marriage of the traditional tropical *'pataugas'* with the design of the leather *'rangers'*. Section and team leaders usually carry binoculars and a 'handie-talkie' radio.

G: Operation 'Léopard'; Kolwezi, Shaba Province, Republic of Zaire, May 1978

G1: Capitaine, 1ᵉʳCompagnie, 2ᵉREP; 19 May

The beginning of the 1970s saw the introduction, as barracks and combat dress, of the new *'tenue de combat Mle 1964'* in olive so-called 'Satin 300' material, most immediately recognisable by the vertically zipped chest pockets. In the Legion it is tightly tailored and elasticated, and the jacket is cut short to the hips. It was in this uniform that the first three rifle companies and the reduced tactical HQ of the regiment jumped over DZ Alpha on 19 May.

The Mle 1956 paratroopers' version of the French Mle 1951 steel helmet has a uniform-cloth cover, a black rubber retaining band, and sub-unit symbols painted on the back: triangles in company colour, with platoon (*section*) numbers. This company commander would have the green triangle of 1ᵉʳCie. and the white 'K' of the HQ platoon.

The jacket displays the usual 'velcro'd' rank tab—here, the three horizontal gold stripes of captain; and, when not tactical, a name tab in black-on-white capitals 'velcro'd' on the right breast. Scarves in company colours are tied round the left shoulder as field signs: 1ᵉʳ—green; 2ᵉ—red; 3ᵉ—black; 4ᵉ—light grey; CCS—yellow; and CEA—blue.

On the left suspender a large 'commando' first aid pack (previously worn on the belt—see B2) contains dressings, burn treatment, morphine syringe, tablets of salt, halazone and dexadrine, and a nerve gas antidote. On the belt are the usual bayonet/fighting knife; the MAC 50 pistol, often secured by a chain lanyard to the belt; the double magazine pouch; and, at the back, two canteens. Some junior leaders also carried the MAT 49 and one or two pouches.

G2: Grenadier-voltigeur, 2ᵉSection, 2ᵉCompagnie, 2ᵉREP; 19 May

The helmet cover (and rucksack—see photo p. 29) were both marked with the company triangle and platoon number, and the company scarf was worn at the left shoulder. Most paras packed very little for the Kolwezi jump: an extra canteen, ration packs, the Legion's green zipped cardigan jacket for the cold African nights, ammunition, and grenades; and the poncho rolled under the rucksack. Even so, many men on the first drop had only four magazines—an incredible 40 rounds!—and this explains why photos show many légionnaires later carrying captured G3 rifles and ammunition. On the belt are the double magazine pouch; the rifle cleaning kit (hidden here) would be in the old 1946 leather pouch, now darkened with boot polish; and the canteen. On the front of the belt this man would wear the bayonet, and a triple grenade pouch tied down to the right thigh.

G3: Tireur d'élite, 4ᵉCompagnie, 2ᵉREP; 21 May

There are three squad snipers in each rifle platoon, armed with the 7.5 mm FR F1—a remanufactured MAS 36 with a removable ten-round magazine, a

The medal most often seen on 2ᵉREP tunics these days: the *Médaille d'Outre-Mer* or 'Overseas Medal', which until June 1962 was the *Médaille Coloniale*. A silver medal on a pale blue and white ribbon, it bears gilt clasps—here, commemorating a tour of duty in Tchad. (Simon Dunstan)

folding bipod, a flash guard/compensator at the muzzle, a pistol grip stock, and the L806 3.8 'scope sight. The 4th Co. dropped with the recce platoon, the mortars, and the remainder of the CCS on the morning of 20 May, and seem to have worn the 'general service' camouflage fatigues from the outset. For one of the motorised fighting patrols through outlying suburbs the sniper would wear minimal equipment: belt, suspenders, magazine and grenade pouches, knife, canteens, and first aid pack. Scarves of parachute material are popular; and this soldier has personally provided himself with tinted ski-goggles.

H: Operation 'Epaulard 1'; Beirut, Lebanon, August–September 1982

H1: Grenadier-voltigeur, Compagnie de Commandement et des Services, 2ᵉREP

H2: Grenadier-voltigeur, 3ᵉCompagnie, 2ᵉREP

During their supervision of the PLO evacuation, and their subsequent patrolling, the légionnaire-paras wore the same fatigues and helmet as at Kolwezi, but now with the new Mle 1974 webbing and the FAMAS rifle. The small FAMAS cleaning pouch is supposed to be worn on the right suspender, but gets in the way, and is often moved to the back of the belt. The rear suspender

A section sniper or *'tireur d'élite'* of the 2ᵉREP contingent in Beirut, August–September 1982, carrying the FR F1 rifle: see commentary, Plate G3. The new Mle 1974 webbing set, introduced in 1980, includes special paired ammunition pouches for the FR F1, smaller than the three-clip pouches for the FAMAS; they can just be made out here, above the canteen. The new webbing retains the triple grenade pouch taped to the right thigh, copied from the old US Army model. The dark strip above the sniper's right chest pocket is the 'velcro' for his name tag, which is removed when 'tactical'. Note the company-coloured triangle and platoon (*'section'*) number marked on the back of his slung helmet.

fastenings are often secured by tying bootlaces through the belt eyelets. Both these paras would wear company-coloured scarves, and national insignia. On the right arm a large olive drab brassard bore the French tricolour; on the left, a small tricolour shield with 'FRANCE' in dark blue on the white top strip was fixed with press-studs or hooks-and-eyes.

H3: Commandant, CCS, 2ᵉREP
Sometimes worn in Beirut with the right sleeve brassard, this is the Legion's summer walking-out and service dress. The pale stone shirt and slacks are worn with a narrow webbing belt copied from the US Army pattern of the Second

World War, with a plain brass buckle plate and tag. Shoulder boards and sleeve *écusson* are worn on the shirt; and the regimental *fourragère* is in 'walking-out' position, its hanging end hooked up, and the ferrule tucked behind the ribbon bar—which is fixed to the shirt with press-studs at each end. The ribbons are the *Ordre National de Mérite*; the *Croix de Valeur Militaire*; and the *Médaille d'Outre-Mer*, 'Overseas Medal', as the old Colonial Medal has been tactfully retitled in recent years! On the right breast below the parachutist's *brevet* the regimental badge hangs on a green leather fob stitched in red. The *képi* conforms to the description under Plate C2.

Note the special shirt creases affected by the Legion: three vertically down the upper chest, one below the pocket, and a double box crease down the outside of the sleeve matching the width of the *écusson*. There are also five vertical and two horizontal creases on the back, making 19 in all—which cursing *légionnaires* have to press in every time the shirt is washed!

I1: Caporal, 2⁰REP, tenue de tradition; Calvi, Corsica, 1982

Troops first received the four-pocket tunic and plain slacks for walking-out dress in c.1960; in 1968 there was a change in material—to the slightly greenish khaki 'Tergal'—but the cut remained the same. Since 1979/80, when the old 1946 battledress was withdrawn, it is also worn for parades, with the embellishments illustrated. (*Errata:* We are informed that the 2⁰REP do not wear white gloves or laces, and wear the *Brevet d'Alpiniste Militaire* low on the left breast pocket.)

All the individual items and insignia worn by this *caporal* in his second engagement are explained in earlier commentaries, apart from the upper badge worn on the right breast (*Brevet d'Alpiniste Militaire*, suggesting that he may belong to the 2⁰Cie.), and the medals. These are the *Médaille d'Outre-Mer*, with clasp '*Liban*'; the 1982 *Médaille de la Défense Nationale*, with clasps '*Mission d'Assistance Extérieur*' and '*Troupes Aéroportées*'; and one of the dozen awards of the Republic of Zaire's *Croix de Bravoure Militaire avec Palme* which President Mobutu presented to the 2⁰REP for what he was pleased to call 'the Second Shaba War' of 1978.

Among the most recent 'semi-official' sub-unit badges produced for 2⁰REP personnel are these. The large gold badge of the *Transmissions* (Signals Platoon) has a red dragon, sword and wing; a green grenade; and blue lettering. The badge of the regimental band, which doubles as the mortar platoon of the support company, is halved green (left) and red; lyre and lettering, silver; grenade, gold; and 'CEA', yellow with silver trim. Other badges known to exist include that of the 4th Co.'s *Section de Sabotage*; a *Groupement d'Instruction de Combat de Nuit*; and commando training and combat swimmer badges. Since many of these sub-unit badges are not officially approved—'*non homologue*'—it is hard to be dogmatic about their actual status.

I2: Sergent, Police Militaire, 2ᵉREP; Calvi, Corsica, 1978
From a photo of regimental patrols rounding up légionnaire-paras for the Kolwezi operation. It is unusual to see these insignia worn on the barracks dress fatigues, which have clearly been 'tarted up' for regimental police use. Rank and name tabs are worn on the chest, and the patch of the 11ᵉ Division Parachutiste on the right shoulder. The NCOs' shoulder boards and the regimental citation lanyard are also worn, together with an MP ('PM') brassard halved in the Legion's green and red. The whitened leather parade belt, with added cross strap, supports an automatic in a whitened leather holster.

The *képi* worn by *sous-officiers* (including, since at least 1961, *caporal-chef*) is in midnight blue and red. Around the top of the body is a single dark blue piping; from this, single lengths of vertical piping rise over the red edge of the crown at the usual four positions; and the top surface is surrounded by a single edging of the same piping. The gold false chin strap is never worn 'down': for parades a second chin strap is added, in black with gold edging, and is worn beneath the chin. The grenade is in gold embroidery. For certain ceremonial occasions *sous-officiers* may wear the white *képi*; e.g. when parading with a colour party.

(The troops' *képi* has been made in white—rather than having a separate white cover—since the early 1970s. There was a period between about 1964 and 1970 when the *képi* was made in pale khaki drill colour, rather than midnight blue and red, though it was still worn with a white cover at all times.)

I3: Grenadier-voltigeur, 2ᵉREP; Corsica, 1984
The Legion para of today, in his element—about to emplane for a jump. Over the standard fatigues and 1974 webbing he wears the latest 'rig'—Individual Parachute Equipment 696–26. The FAMAS is now carried in a purpose-designed EL 33 weapon case hitched to the left side of the special cover, EL 32, containing the rucksack and other equipment. The helmet, with the same type of cover and band as the old Mle 1956, is the F1 (Mle 1978). This appeared in Legion service during 1984. The shell—which is made in one size, with a sophisticated system of adjusting interior strapping—has a distinctly Russian appearance when seen uncovered and from the side, with deep extensions over the ears. The three-point straps and chin cup are worn by all personnel, not only airborne troops.

J, K, L: Insignia
These plates show all official unit insignia worn by Legion parachute units to date; but it is possible that there may be further changes among the semi-official 2ᵉREP company badges by the time of publication. They are reproduced close to actual size, though we do not claim accuracy to the millimetre. (Collectors are strongly recommended to acquire a copy of *Les Insignes de la Légion Étrangère* by Christian Malcros, published by the Legion's Information and Historical Service and available from IILE, 13114 Puyloubier, France. It includes badges up to the mid-1970s, with full accompanying text to the 250-plus colour photographs.)

J background: US camouflage-printed material as worn by the Legion's BEPs in Indochina.
J1: Compagnie Parachutiste du 3ᵉRégiment Étranger d'Infanterie, April 1948–May 1949
J2: 1ᵉʳBataillon Étranger de Parachutistes, July 1948–December 1950 and March 1951–September 1955
When the unit became the 1ᵉʳRégiment Étranger de Parachutistes in September 1955 this badge was retained, with the sole difference that the 'B' lost its bottom bar to become 'R'. The modified badge was worn until the unit's disbandment in April 1961.
J3: 2ᵉBataillon Étranger de Parachutistes, October 1948–December 1955; and 2ᵉ Régiment Étranger de Parachutistes, December 1955 to present
J4: 3ᵉBataillon Étranger de Parachutistes, November 1949–September 1955; and 3ᵉRégiment Étranger de Parachutistes, September–November 1955
J5: 1ᵉʳᵉCompagnie Étrangère Parachutiste de Mortiers Lourds, September 1953–June 1954
J6: Compagnie Étrangère de Ravitaillement par Air, January–August 1951 During this brief period the air resupply despatch company operating out of the Base Aéroportée Nord at Hanoi/Gia Lam was specifically affiliated to the Legion (and largely composed of former personnel of 3ᵉ Cie., 2ᵉBEP).
J7: 1ᵉʳᵉCompagnie, 2ᵉREP, mid-1970s to present During 1974/75 the separate companies of the regiment began to acquire unofficial insignia reflecting their special company rôles. (These are not

tolerated on formal uniform.) This design obviously refers to the company's night-fighting mission.

J8: 2ᵉCompagnie, 2ᵉREP, first type, 1975 The design recalls several deployments through Djibouti, symbolised by the antelope head.

K background: British camouflage-printed material as worn by the Legion's BEPs in Indochina.

K1: 2ᵉCompagnie, 2ᵉREP, second type, 1977.

K2: 2ᵉCompagnie, 2ᵉREP, third type, 1982 The latest series of company badges are all round or oval.

K3: 3ᵉCompagnie, 2ᵉREP, first type, 1974 Another reference to service in Djibouti, whose map outline appears here with the acronym of the territory's pre-independence title: 'Territoire Français des Afars et des Issas'.

K4: 3ᵉCompagnie, 2ᵉREP, second type, 1982.

K5: 4ᵉCompagnie, 2ᵉREP, 1975.

K6: Compagnie de Commandement et des Services, 2ᵉREP, 1982.

K7: Compagnie d'Éclairage et Appui, 2ᵉREP, 1982.

K8: Fourragère worn by the 1ᵉʳBEP and 1ᵉʳREP, 1953–61 The end of the citation lanyard, in the yellow and green of the *Médaille Militaire*, worn by the unit after its fourth collective citation on 17 December 1953. (*Errata:* The lanyard apparently did *not* bear the TOE *'olive'* illustrated.)

K9: Fourragère worn by the 2ᵉBEP and 2ᵉREP, 1954 to present The sixth collective citation of a French Army unit brings the lanyard in the colour of the *Légion d'Honneur*; the 2ᵉBEP received its sixth citation on 23 April 1954, while fighting inside Dien Bien Phu. The lanyard ferrule end bears the additional *'olive'* in the colours of the *Croix de Guerre TOE*, displayed on lanyards awarded for fourth and sixth citations in Indochina.

L background: French camouflage-printed material as worn by the Legion's BEPs and REPs in Indochina and Algeria.

L1: Shoulder patch, 10ᵉ Division Parachutiste, July 1955–April 1961 Worn on the right shoulder by the 1ᵉʳREP, for parade, walking-out, and occasionally for operations. These patches were usually sewn to a black background, and fixed with press-studs or hooks-and-eyes. The colours recall the original beret colours of the three

Légionnaire Rob Flynn, wearing the new *tenue de parade* which was issued to the 2ᵉREP in 1985. A modification of the four-pocket tunic, it resembles the old battledress blouse, but is lighter weight. The skirt and lower pockets are removed, and box pleats and a buttoned cuff band are added to the sleeves. This remodelled garment was first worn by the bandsmen of the Legion's *Musique Principale*. (Photo via Mike Miles)

categories of French parachute units—blue for 'Metropolitan' Chasseurs Parachutistes; red for Parachutistes Coloniaux; and green for Parachutistes Étrangers.

L2: Shoulder patch, 25ᵉ Division Parachutiste, June 1956–April 1961 Worn by the 2ᵉREP; again, the design incorporates blue, red and green.

L3: Shoulder patch, 11ᵉDivision Légère d'Intervention, May 1961–December 1963 The 2ᵉREP did not serve with this formation upon its creation, thus the absence of Legion green. The design incorporates the eagle of the old 10ᵉ and the parachute of the old 25ᵉDivision. In December 1963 the formation was renamed simply '11ᵉ Division'. By mid-1967 the 2ᵉREP was integrated into the 11ᵉDivision.

L4: Shoulder patch, 11ᵉDivision Parachutiste, April 1971 to present The change of title in April 1971 brought a new patch, this time incorporating Legion green once again.

Notes sur les planches en couleur

A1 Casque américain M1; survêtement de camouflage britannique 1942; pantalon vert français 1947; équipement de toile américain et bottes à agrafes de type américain. Il porte le fusil CR.39 et la 'havresac spéciale' pour chargeurs FM 24/29. **A2** Veste camouflée américaine 1943 et pantalon camouflé britannique 1942—tenue de combat normale des BEP; casque, équipement de toile américain et 'musette spéciale' 1924 contenant six chargeurs. **A3** Tenue de combat vert olive 1947; chapeau de brousse français; équipement de toile américain et carabine M1A1. Notez l'écusson et les insignes de grade portés non sur l'épaule mais sur la poitrine.

B1 Veste camouflée américaine avec motif original 'USMC', écusson et chevrons portés sur la poitrine. Ceinture et cartouchière en cuir de l'équipement français de 1946, portés en de rares occasions par les parachutistes; PM MAT 49. L'arme à l'avant plan est le 'recoilless rifle' M18 57 mm. **B2** Tenue coloniale Mle 1951, parfois portée par le BEP. Le devant de cette tenue ressemblait à celui de B3 mais elle ne comportait pas les petites poches sur les cuisses. **B3** Tenue camouflée TAP Mle 48/51 ressemblant à la tenue verte portée à B2 mais sans les poches sur les pans arrière de la veste. Béret vert, autorisé pour les BEP en 1949. Copie française de l'équipement US de toile, 'Mle 1950 TAP'.

C1 Uniforme de parade d'été standard de la Légion; le béret blanc n'étant porté pour la parade que par le personnel indochinois. Fourragère Croix de Guerre TOE décerné au bataillon par la seconde citation du bataillon, 25.1.1951. **C2** Mélange des uniformes de parade d'été et d'hiver des officiers. La fanion est d'après une photo; l'exemplaire survivant à Aubagne est d'un modèle différent. **C3** Uniforme d'hiver 'battledress' 1946 avec embellissements de parade de la Légion: épaulettes, ceinture bleue et, ici, ceinture blanchie. Képi recouvert en blanc porté par tous les grades au-dessus de caporal-chef.

D1 Tenue camouflée TAP 1956 avec poches de rebord posées droites et non obliques, et sans la 'queue' et les petites poches de cuisse. Les galons de grade étaient portés sur la poitrine ou sur les épaulettes par les officiers et les adjudants. Equipment léger pour service de sécurité urbaine; équipement de toile français 1950, avec pistolet MAC 50 et PM MAT 49. Insigne du 10e DP portés en de rares occasions. **D2** Veste matelassée d'hiver et toque, souvent portés an Algérie. **D3** Notez le rucsac 'bergen', les bottes 'rangers' françaises, le fusil MAS 49/56 et les cartouches de la mitraillette légère de la groupe. Le chèche est noué dans un style typique.

E1 Typiques de l'époque, insignes et médales portés sur l'uniforme de camouflage pour les parades. **E2** Combinaison inhabituelle, d'après une photo, du béret et de l'uniforme de sortie d'hiver. **E3-8** Insignes de grade sur la manche et l'épaule—par suite du manque de place, voir les légendes en anglais.

F1 Tenue de combat vert olive de parachutiste, portée à partir de 1966 environ; l'uniforme de camouflage illustré à G3 était aussi porté au Tchad. Notez la mitraillette légère AA 52. **F2** Veste de camouflage sans caractéristiques 'parachutiste'—c'est-à-dire qu'elle possède des boutons dissimulés sur le devant, les poches et les manches et un col chemise—portée avec des shorts verts ou sable à Djibouti. Les deux diagonales or indiquent le rang de sergent. **F3** Gilet de combat et shorts; nouvel équipement de toile 1974; fusil FAMAS remplaçant la MAS 49/56 et la MAS 49; galons de grade—un or, deux vert—de caporal-chef.

G1, G2 Les premières troupes parachutées à Kolwesi portaient cet uniforme de combat 1963, en tissu 'Satin 300'; les deuxièmes, voir G3—de tenues de camouflage. Les foulards d'épaule permettaient de reconnaître la compagnie: 1ère, vert; 2ème, rouge; 3ème, noir; 4ème, fis; CCS, jaune; et CEA, bleu. Le triangle à l'arrière du casque dans la couleur de la compagnie, portant le numéro de section ou 'K' pour le quartier général; cela était répété sur le 'bergen'. **G3** Chaque groupe possèdait un tirailleur équipé du fusil FR F.1; pour une patrouille en dehors de Kolwesi, cet homme ne porte qu'un équipement léger.

H1, H2 Même uniforme qu'à l'illustration G, mais avec équipement de toile de la Mle 1974 et le fusil FAMAS. Insignes nationaux spéciaux portés sur les deux manches pour les opérations de Beyrouth. **H3** A part le brassard de Beyrouth, uniforme normal de sortie et de parade d'été.

I1 Depuis 1980, la tunique de sortie à quatre poches est aussi portée pour les parades, remplaçant la 'battledress' de 1946. **I2** Combinaison inhabituelle de tenue de combat et de nombreuses insignes, portée par la police du régiment—d'après une photo. Notez le style 'sous-officier' du képi. **I3** Le nouveau casque 1978; équipement individuel de parachute 696-26, avec gaine EL 32 pour l'équipement et étui EL 33 pour fusil.

J, K, L Les insignes sont clairement identifiés dans les légendes en anglais. L'illustration J présente en arrière-plan le tissu de camouflage américain porté en Indochina; K, tissu de camouflage britannique porté en Indochine; L, tissu de camouflage français de la Mle 1951, porté à partir d'environ 1953 à 1966.

Farbtafeln

A1 Amerikanischer M1 Helm; britischer Tarnkittel von 1942; grüne französische Hosen von 1947; amerikanisches Gurtwerk und Schnallenstiefel im amerikanischen Stil. Er trägt ein CR.39 Gewehr und den Spezialbeutel für FM 24/29 Magazine. **A2** Amerikanische Tarnjacke von 1943 und britische Tarnhosen von 1942, die normale Kampfausstattung für die BEPs; amerikaner Helm, Gurtwerk und Stiefel; FM 24/29 und Spezial-Musette von 1924 mit sechs Magazinen. **A3** Französische olivgrüne Kampfbekleidung von 1947; französische Buschmütze; amerikanisches Gurtwerk und M1A1 Karabiner. Man beachte das écusson und Rangabzeichen, die nicht auf den Ärmeln, sondern auf der Brust getragen werden.

B1 Amerikanische Tarnjacke mit originalem USMC Aufdruck und écusson und Abzeichen auf der Brust. Ledergürtel und Beutel aus der französischen Ausrüstung von 1946, gelegentlich von Fallschirmjägern getragen; MAT 49 Maschinenpistole. Die Waffe im Vordergrund ist ein 57 mm M18 'recoilless rifle'. **B2** Tenue coloniale Mle 1951, gelegentlich von BEPs getragen. Die Vorderseite dieses Anzugs ist ähnlich wie auf B3, aber ohne die kleinen Taschen auf den Oberseiten der Schenkel. **B3** Tenue camouflée TAP Mle 48/51, ähnlich dem grünen Anzug auf B2, aber ohne die Taschen in den hinteren Jackenschössen. Grünes Barrett, 1949 für die BEPs zugelassen. Französische Kopie des amerikanischen Gurtwerks, 'Mle 1950 TAP'.

C1 Standard-Sommerparadeuniform der Fremdenlegion mit nur von Mitgliedern in Indochina getragenem weissen Barrett. Fourragère des Croix de Guerre TOE wurde anlässlich der zweiten ehrenvollen Erwähnung des Bataillons am 25.1.1951 verliehen. **C2** Mischung aus Sommer- und Winterausführung der Paradeuniform für Offiziere. Der fanion stammt von einem Photo; das erhaltene Exemplar in Aubagne hat ein anderes Design. **C3** 'Battledress' Winteruniform von 1946, mit Paradeschmuckabzeichen der Fremdenlegion: Epauletten, ceinture bleue und in diesem Fall weisser Gürtel. Ränge unterhalb des caporal-chef trugen einen weiss bespannten képi.

D1 Tenue camouflée TAP von 1956 mit geraden statt schräg angesetzten Rocktaschen und ohne die 'Schösse' und die vorderen kleinen Oberschenkeltaschen. Rangabzeichen wurden von Offizieren und adjutants auf der Brust oder auf Schulterstreifen getragen. Leichte Ausrüstung für städtischen Sicherheitsdienst: französisches Gurtwerk von 1950 mit MAC 50 Pistole und MAT 49 Maschinenpistole. Divisionsabzeichen der 10e DP wurden nur gelegentlich getragen. **D2** Wintersteppjacke und 'toque', häufig in Algerien getragen. **D3** Man beachte den bergen-Rucksack, die französischen rangers-Stiefel, das MAS 49/56 Gewehr und die Patronen für die leichten MG der Staffel. Der Schal (cheich) ist auf typische Weise geknotet.

E1 Zeichen und Auszeichnungen werden, typisch für die Periode, bei Paraden auf Tarnuniform getragen. **E2** Ungewöhnliche Kombination, nach einem Photo, von Barrett und formeller Winter-Ausgehuniform. **E3-8**: Ärmel- und Schulterrangabzeichen (aus Platzgründen siehe bitte englische Bildunterschriften).

F1 Olivgrüne Kampfuniform für Fallschirmjäger, ab etwa 1966 getragen; die in G3 abgebildete Tarnuniform wurde auch im Tschad getragen. Man beachte das leichte AA 52 MG. **F2** Tarnjacke ohne Fallschirmjäger-Kennzeichen, d.h. mit verdeckten Knöpfen vorne, an den Taschen und an den Manschetten, und mit vollem Kragen, in Dschibuti mit grünen oder sandfarbenen Sandalen getragen. Das Rangabzeichen (zwei goldene Diagonalen) identifiziert einen sergent. **F3** Kampfweste und Shorts; neues Gurtwerk von 1974; FAMAS Gewehr anstelle der MAS 49/50 und MAT 49; Rangabzeichen (eines gold, das andere grün) eines caporal-chef.

G1, G2 Die erste Absprunggruppe trug in Kolwesi Kampfuniformen 1963 aus Satin 300, die zweite (siehe G3) Tarnuniformen. Schulterschals identifizieren die Kompanie: 1. grün, 2. rot, 3. schwarz, 4. grau, CCS gelb und CEA blau. Dreieck hinten auf dem Helm in Kompaniefarbe mit der Zugnummer oder K für das HQ. Diese Markierung wird auch auf dem Rucksack wiederholt. **G3** Jede Gruppe hat einen Scharfschützen mit FR F.1 Gewehr; für eine Patrouille ausserhalb von Kolwesi trägt dieser Mann nur eine leichte Ausrüstung.

H1, H2 Gleiche Uniform wie auf Tafel G, aber mit Mle 1974 Gurtwerk und FAMAS Gewehr. Spezielle Landesabzeichen auf beiden Ärmeln für Einsätze in Beirut. **H3** Abgesehen von der Armbinde für Beirut eine normale Sommerparade- und Ausgehuniform.

I1 Seit 1980 wird die Ausgehjacke mit vier Taschen auch bei Paraden getragen und hat das 'battledress' von 1946 ersetzt. **I2** Ungewöhnliche Kombination von Kampfbekleidung und zahlreichen Abzeichen, von der Regimentspolizei getragen; nach einem Photo. Man beachte den képi im Stil eines sous-officier. **I3** Der neue Helm von 1978; équipement individuel de parachutiste 696-26, mit EL 32 Behälter für die Ausrüstung und EL 33 für das Gewehr.

J, K, L Die Abzeichen sind in den englischen Bildunterschriften identifiziert. Tafel J hat einen Hintergrund von amerikanischem Tarnmaterial, wie es in Indochina getragen wurde; K: britisches Tarnmaterial, ebenfalls in Indochina getragen; L: französisches Mle 1951 Tarnmaterial, von ca. 1953-66 getragen.